Latin American Development: Geographical Perspectives

Latin American Development:
Geographical Perspectives

Edited by David Preston

LIVERPOOL
UNIVERSITY
LIBRARY

Copublished in the United States with
John Wiley & Sons, Inc., New York

Longman Scientific & Technical,
Longman Group UK Limited,
Longman House, Burnt Mill, Harlow,
Essex CM20 2JE, England
and Associated Companies throughout the world.

Copublished in the United States with
John Wiley & Sons, Inc., 605 Third Avenue, New York 10158

First published 1987

British Library Cataloguing in Publication Data
Latin American development: geographical
  perspectives.
  1. Anthropo-geography – Latin America 2. Latin
  America – Historical geography
  I. Preston, David A.
  304.2'098   GF514
ISBN 0-582-30148-3

Library of Congress Cataloging-in-Publication Data
Latin American development.

  Bibliography: p.
  Includes index.
  1. Latin America – Economic conditions. 2. Latin
America – History. I. Preston, David, 1936–
HC123.L38   1987   338.98   86–7457
ISBN 0–470–20783–3 (Wiley, USA only).

Set in Linotron 202 10/11 pt Palatino
Produced by Longman Group (FE) Limited
Printed in Hong Kong

Contents

List of Figures

List of Plates

List of Contributors

Dr John Dickenson, Senior Lecturer in Geography, University of Liverpool

Dr Alan Gilbert, Reader in Geography, University College and Institute of Latin American Studies, London

Dr Robert N. Gwynne, Lecturer in Geography, University of Birmingham

Dr Arthur Morris, Senior Lecturer in Geography, University of Glasgow

Dr Linda Newson, Lecturer in Geography, King's College, London

Dr David Preston, Senior Research Fellow in Human Geography, Australian National University, Canberra

Dr Richard Smith, Senior Lecturer in Geography, University of Leeds

Dr Janet Townsend, Lecturer in Geography, University of Durham

Introduction

This book seeks to report on the state of knowledge about selected
aspects of Latin America. As geographers we share a concern for
understanding the spatial patterns of change in Latin America both
now and in the past but the views on which we report are those of
a range of social scientists, including geographers, who have
grappled with the major problems of interpreting Latin America.
These essays seek to identify the main issues that concern us and
Latin Americans in the attempt to understand more completely past
and present patterns of change across the continent. To guide the
reader further each author has included at the end of each chapter a
short list of published work in English, that will enable further
exploration of the topics discussed. The interested reader who
intends to visit Latin America (even from an arm chair) should
consult the latest edition of the annual *South American Handbook*
(obtainable at good bookshops or direct from Trade and Travel
Publications, Parsonage Lane, Bath BA1 1EN, UK). The British
reader keen to find out about intellectual frontiers currently being
crossed should consider joining the Society for Latin American
Studies. This has two publications: a termly Newsletter which lists
lectures, seminars and exhibitions throughout the country and the
Bulletin of Latin American Research, published twice yearly, which
contains scholarly articles about recent research. Most interesting to
the newcomer to Latin American studies would be the Society's
Annual Conference to which visitors are always welcome. Details
about the Society can be obtained by writing to the Membership
Secretary, SLAS, c/o Centre for Latin American Studies, The
University, Liverpool L69 3BX, UK. North American readers should
consider joining the Latin American Studies Association which
publishes a quarterly journal, *Latin American Research Review*
containing articles, comments and book reviews and which holds a
major meeting every eighteen months. For details write to the
Association at Sid Richardson Hall, University of Texas, Austin Tx
78712-1284, USA.

The present volume is one of several volumes being prepared at

the initiative of the Developing Areas Research Group of the
Institute of British Geographers where seminars and meetings during
the past ten years have seen the airing of many of the ideas that are
presented here. We gladly acknowledge the initiative of this Group
and especially of Denis Dwyer, a past Chairman of the Group, who
artfully provided the nest into which we have laid these literary
eggs, guaranteed free-range and of varied colours appropriate to our
different views. As editor of this volume I am very grateful for the
rapidity with which my colleagues wrote and even revised their
chapters leaving me toiling to catch up and finally conclude the
volume while preparing to explore Spain's former outpost in the
Pacific – the Philippines – from an Antipodean haven.

David Preston
January 1986

Acknowledgements

We are grateful to the following for permission to reproduce copyright material:

Croom Helm Ltd for figs 5.2, 5.3, 5.5, 5.6 (R. N. Gwynne, 1985), Tables 5.7, 5.9 (R. N. Gwynne, 1982); the Author, Peter Evans for Tables 5.5, 5.6 (G. Gereffi & P. Evans, 1981); The Johns Hopkins University Press for Table 8.2 from Table 2.8, p 46 (G. P. Kutcher & P. L. Scandizzo, 1981), Table 8.3 from Table 12, p 82 (O. Altimir, 1982); Westview Press Inc. for Table 8.5 from fig 6.8 (S. Hecht, 1983); John Wiley & Sons Ltd for Table 5.8 (J. A. S. Ternant, 1976).

CHAPTER 1

Latin America in the world

David Preston

Latin America can claim leadership in the world in various ways. It contains, without doubt, the most advanced group of Third World nations in terms of economic development. The capacity for industrial production of goods, previously the monopoly of the northern industrial nations, in some Latin American nations – notably Brazil – is a sign of entrepreneurial skill and technical competence. In intellectual spheres Latin Americans have made notable contributions in the social sciences, in mathematics and especially in the cinema. Spanish-American prose and poetry is admired and imitated in Spain and is a major influence in literary movements elsewhere.

We are not dealing with a group of extremely poor, backward nations, each struggling to achieve impossibly high levels of living based on limited attainments. The most striking feature about Latin America, as with many Third World areas, is the contrasts that exist both between and within countries. European visitors are invariably impressed by the juxtaposition of extremes of poverty and conspicuous affluence that each characterize sizeable sections of Latin American metropolitan populations and are unsurpassed in western Europe. To those who visit Latin America regularly the spirit of dynamism and the ready acceptance of change are exhilarating after the more turgid attitudes to, and limited expectations of, change prevailing in the Old World.

The native population of Latin America, when the Europeans arrived in the 1500s, was largely organized into a variety of territorial units, ranging from simple hunter-gatherer societies to the complex Aztec and Inca states. The degree of inter-regional contact was limited, although there was undoubtedly some long-distance trading between Central and South America and sporadic trans-Pacific contact also took place. The creation of the Spanish Empire, and later, but to a lesser extent, the Portuguese Empire, in the Americas, for the first time, brought a major continental area under political control of European nations and provided Europe with a source of minerals and other products that stimulated trade and

1

industry as a whole. The New World had become fully involved in the web of economic and political relations that Asian nations managed to evade so successfully for several centuries.

While India, the East Indies and China became increasingly influenced by the importance of trade with the European powers from the seventeenth century onwards, this was accompanied not by political control from Europe, but rather by the establishment of trading outposts. Even during the eighteenth and early nineteenth centuries the manner of imperial domination by the British in India or the Dutch in the East Indies was less absolute or geographically comprehensive than comparable Spanish and Portuguese domination of Latin America. A major factor which may account for this is the much larger populations in India and the East Indies and the higher level of economic and political development that obtained in many parts of the Far East in comparison with the New World, although Iberian colonialism was much concerned with Christianizing the native population and, as such, affected native society profoundly.

Although the founding of towns was a major part of the Spanish imperial system of control, the supply of precious metals was the main contribution of Latin America to the European economy in the early period of European colonial influence. Within a century, a range of commodities were also exported – sugar, cacao, indigo and cochineal, for example – which, as Linda Newson describes in the next chapter, satisfied European urban demand for luxury goods. Where agricultural enterprises were established to produce specific commodities, using forced labour, the basis of one new and peculiarly New World socio-economic unit – the plantation – was established. Nothing similar and on such a scale was attempted elsewhere in the territories of the European colonial powers until the nineteenth century.

The contribution of Latin America to the rest of the world as a consequence of being drawn into the European imperial economic system was more than just silver and other precious metals. Crops introduced into Europe from America – particularly maize and potatoes – spread far, reaching eastern Asia within a century of the discovery of the Americas. While Amerindian agriculture was being transformed by the introduction of European livestock – largely sheep and cattle – and bread grains, so was agriculture in some parts of Europe changing to cultivate the new American crops. This was to be the first major introduction of new crops to Europe since the Arabs had brought their farming systems to Spain many centuries earlier. Trade with the New World by the Europeans by the late eighteenth century was proving more profitable than with anywhere else in the world: Latin America was sending back to Europe every year four or five times as much as India, which contained five times as many people (Braudel 1984).

In the following chapters Linda Newson, Richard Smith and John Dickenson each trace the impact of the period of colonial contact on society, economy and land in Latin America. Unlike any other major world area, Central and South America were completely dominated by alien powers for nearly three centuries. The metropolitan nations shaped the manner of growth and change in such a way that subsequent developments strongly reflect, within states and across the continent, a common colonial heritage. By contrast in India, the East Indies, China and West Africa, European influences were apparent only in certain locations and related to the production of and trade in certain commodities. In Europe, those who benefited from the colonization of Latin America were not located solely in the Iberian peninsula. The volume of trade to and from Latin America was such that other commercial and maritime powers were involved both legitimately, as contraband traders, and through piracy. Much of the trade in goods coming from and destined for Latin America originated in other European countries. This trade was controlled first by the Netherlands (until 1730) and then by Great Britain, making the exploitation of Latin America during the centuries following Columbus' voyages very important to much of western Europe.

Various of the chapters in this book discuss the path along which change has led Latin America's use of space in the period since Independence. In this, in identifying the distinctive nature of Latin America's development in the contemporary world the importance of autochthonous change has not been overlooked. Many historical accounts of the continent are by Europeans and stress the importance of European, and Anglo-American influences on ideas in the southern continent. Yet at a local and regional level scholars, political leaders and ecclesiastics discussed major regional and national issues, often in the light of events and ideas developing elsewhere in the world and directed local events towards a destiny that was relevant in the interests of the people of Valencia, Tucuman or Cuzco. While the many changes taking place throughout the continent at different periods may be interpreted in terms of world historical patterns, many of those responsible for implementing change were primarily concerned with their own region or nation.

During the nineteenth century, Latin America became politically independent but the commercial dominance of Spain and Portugal was replaced by that of Great Britain which was the dominant overseas commercial partner of much of the continent until the middle of the twentieth century. United States' influence, at first concentrated in the Spanish Caribbean, Central America and Mexico, grew to encompass all of Latin America later in the twentieth century. The significance to Latin America of the new form of domination was that it was primarily commercial with political control imposed in order to maintain freedom to control commercial activities. The geographical importance of this new form of domination was that the areas of the continent that benefited most from the various activities were peripheral to the main population centres. As John Dickenson indicates in Chapter 4, the nature of British investment was far more in mining and agriculture and the related infrastructure of highways, railways, banking and later telegraph and radio communications, than it was in stimulating the already populated and productive regions. Such development did benefit the big cities that served as commercial, and political centres. It was here that governments needed to be influenced and where financial institutions were logically located. Cities, such as Buenos Aires, Montevideo and Rio de Janeiro, which combined their capital city function with that of a major port, obtained a double benefit as political and commercial centres and grew accordingly. The Latin American element in this neo-colonial situation was in providing a cadre of upper class managers (in the army, the government bureaucracy and in national political life) whose fortunes were in large measure tied to overseas interests. They and their families were educated in Europe or the USA, maintained homes and investments abroad and formed what, in many respects, was an alien superstructure to national economic and social life. Such were the Trujillos in the Dominican Republic, the Somozas in Nicaragua or the Aramayos and Patiños in Bolivia.

The difference in the paths followed by the British colonies in North America after 1783 and the Iberian colonies in South America after the 1820s is striking. On the one hand Ibero-America continued on the margins of the world economy, as a provider rather than a beneficiary controlled by those powers who dominated the European economy of the time. In the United States by contrast, although the South provided goods to Europe, independent trade and production developed from New England to the Caribbean and over the whole North Atlantic. As a result, although much of American profit was lodged in England, American independent commercial prosperity was seen in the last half of the eighteenth century as a threat to English commercial dominance. No such situation existed in Latin America. Even in the River Plate, extensive legal and illegal trade in the eighteenth century was carried on by European vessels with Buenos Aires serving as an entrepôt rather than as a commercial and manufacturing centre.

Latin American autonomy has taken a different form during the middle of the twentieth century. With the waning of British dominance, the different forms in which other nations have sought to manipulate Latin America for their own ends have allowed internally-generated growth to develop in a variety of ways. This was particularly associated with major currents of migration which enabled large and fertile new regions to be populated and to develop, as I have indicated in Chapter 9. Although export surpluses were managed by the dominant commercial powers, local urban development in particular was based on the provision of goods and services to the agro- exporting industry. The expansion in São Paulo State, for example, of more intensive forms of agriculture stimulated considerable urban and industrial growth. This in turn generated its own wave of growth as a consequence of the demand for goods for the exploding urban market and contributed significantly to the post-

1950 metropolitan expansion. Thus the majority of the meat produced now in western São Paulo, on the fringe of the Gran Chaco is destined not so much for the hamburger chains of North America as for the supermarkets (and hamburger stands) of the São Paulo conurbation. In this sense a great deal of expansion is now responding to Latin American demand rather than to a world demand.

The increasing intellectual independence shown by Latin America is highlighted by three major national revolutions that have occurred in Mexico, Cuba and Nicaragua. Each of these revolutions involved a national search for a new political identity that would more adequately respond to the needs of the population as a whole. The Mexican Revolution was important at a continental level not because of its radical nature, for that was soon diluted, but more because it became a truly national revolution ousting an old-style dictator who had shown little concern for involving the population in the progress that seemed possible at the turn of the century. The Mexican Revolution predated the Russian Revolution, but its impact on the continent was diminished by the prolonged civil war that accompanied it and by the death or reduced influence of the more radical of its leaders.

By 1959 the time was ripe for experiment with more radical reform and the tactical success of the Cuban Revolution concentrated attention on the possibility of there being a Latin American path to socialism free from undue external influence. Despite the vicissitudes of Cuban economic post-revolutionary fortunes and the fact that US hostility compelled Cuba to search for alternative sources of political and economic support, the considerable advances in health and education, in the distribution of wealth and in overall agricultural production have shown that Latin American political movements are not all subverted and do not necessarily end in military dictatorships. A wave of popular activism directed against authoritarian governments involving guerila warfare, considerable loss of life and disruption in different Latin American countries shows the degree of ideological commitment that young Latin Americans have to seeking a solution to their own problems. This far outweighs the negative consequences of the failure of many such campaigns to gain popular support. The success of the broad front of popular support for the Sandinista movement in Nicaragua in the late 1970s further demonstrated that a popular movement can generate guerila warfare capable of overcoming a regular army. It also shows that any such popular movement is as likely to generate as unreserved a hostility from the USA as did the Cuban revolutionary movement twenty years before.

Two relatively small states have shown themselves capable of radically revising their method of government and path towards progress and the experiences of each have aroused considerable interest in Europe and elsewhere in the Third World. Not only the practical example, but also the writings of Latin American and other

5

radical philosophers, with a deep commitment to Latin American autonomous development, have a considerable impact in the Third World, particularly in those countries seeking alternative paths of change to those recommended by partisan international agencies. The ferment of ideas and opinions that leads to the frequent closure of Latin American universities has often caused distinguished Latin American scholars to live for a period in Europe, and occasionally in the USA, where their writings and teaching have had a direct impact on scholarship and thought.

Latin America is more influential than ever before in the world of ideas and in the form of practical solutions adopted to overcome current world problems. Latin America in the next century will undoubtedly play an even more important role among Third World nations particularly if the new generation of popularly elected leaders can continue to demonstrate their capacity to explore new solutions to old problems. The rapidity of city growth across the region means that problems of megalopolitan sprawl will have to be tackled in Brazil, Mexico and Argentina before the century ends, on a scale not yet encountered elsewhere in the Third World. The increasing disparities in levels of living to which Arthur Morris refers guardedly in Chapter 6 must be diminished if the poor are not to seek a revolutionary solution. Military reformist as well as socialist governments have shown a willingness to tackle such problems by dramatic literacy programmes, and by various styles of land reform, but these moves run counter to the ideology of successive US governments, who continue to ignore the counsels of their diplomats and scholars who have lived and worked in Latin America. Thus, any approach to policy-making that seeks new alliances or which seeks to reduce the power of the military seems doomed to be undermined by the US even when such new policies are supported by a variety of First World governments. One way to counter such misinformed intervention, is to inform more persuasively those who seek to know more about the continent, which is the aim of this book.

Further reading

Braudel F (1984) *The Perspective of the World, Civilisation and Capitalism, 15th–18th Century*. Collins, London.

CHAPTER 2

The Latin American colonial experience

Linda Newson

Columbus did not discover America, for the continent had been
settled for over 30,000 years, when he landed in Hispaniola in 1492.
Rather he opened the window on a world that was inhabited by up
to 100 million people who comprised societies ranging from highly
developed agricultural states to simple hunting, fishing and
gathering bands. The pre-Columbian period had been one of
increasing cultural diversity, whereas the colonial period that
followed was one of cultural convergence that gave Latin America its
distinctive character. Nevertheless, during the colonial period,
regional variations existed which reflected not only the priorities and
policies of the colonial powers, but also the continent's pre-
Columbian heritage. As such, a knowledge of the distribution of
Indian cultures and population on the eve of Spanish conquest is
just as important for understanding developments during the
colonial period, as an awareness of the cultural environment in
Europe that led to the conquest of the New World.

The pre-Columbian heritage

The original inhabitants of Latin America were hunters and gatherers
who arrived from Asia having crossed the Bering Strait about 30,000
years ago. Migrating south through North America, they arrived in
Mexico about 20,000 years ago, and following the highlands of
Middle America and the Andes, they reached the Straits of Magellan
about 9000 BC. These early inhabitants of the sub-continent were
hunters of large, now extinct, animals such as mastodons, giant
sloths and horses. As the climate changed at the end of the last Ice
Age and the big game were overexploited, the Indians began to
hunt small game and collect wild vegetables, fruits and nuts, whilst
those who settled on the rivers and coasts fished and collected
shellfish. All of these groups were nomadic or semi-nomadic, living
in temporary shelters, which they abandoned as the food resources
they exploited were depleted. They formed egalitarian social groups,

whose size generally did not exceed thirty persons. Whilst these hunting, fishing and gathering bands originally inhabited the greater part of the Latin American sub-continent, as the Indians turned to agriculture for their subsistence, their distribution became more restricted. By the time of the European conquest they were to be found in remote areas which in general were unsuited for cultivation. These included the cold and wet regions of Tierra del Fuego, where the Indians subsisted on shellfish gathering, and the dry savanna and temperate grassland areas of Brazil, Uruguay and Argentina, where they hunted guanaco (a wild relative of the llama) and rhea (a variety of ostrich), and collected wild vegetable products. Hunters and collectors also existed in the arid north of Mexico, whilst nomadic fishers inhabited the swampy regions of the Orinoco basin.

The process of plant domestication began in Mexico and the Andean area about 7000 BC (see Ch. 3), but several thousand years elapsed before agriculture in those areas formed the basis of Indian subsistence. The development of agriculture provided Indian groups with a more reliable and productive source of food, enabling them to become sedentary and permitting larger population densities to be supported. Although permanent villages had populations ranging up to several thousands in size, the societies were integrated by kinship ties and they remained essentially egalitarian. Initially the type of agriculture that was practised was land-extensive, involving the periodic abandonment of plots to fallow to allow the soil to recover some of its lost fertility. In Mexico and Central America this type of cultivation, known as *milpa* cultivation, was associated with the raising of maize, beans and squashes, and, supplemented by various forms of intensive cultivation, it supported the large Maya population. In lowland South America, where it is known as *conuco* cultivation, manioc and sweet potatoes were the major crops raised. Since root crops are deficient in protein, in the latter area the Indians continued to hunt and fish. There cultivation was practised on the alluvial banks of rivers where fish and game were more abundant and where the soils were naturally more fertile. At the time of the Iberian conquest tribal groups practising *conuco* cultivation were concentrated along the banks of rivers throughout Amazonia, whereas the interfluvial areas tended to be inhabited by hunters and gatherers or else they were unoccupied.

Whilst shifting cultivation, involving fallow periods of variable length, was probably the most common form of cultivation practised in Latin America at the time of the Iberian conquest, in many areas new techniques had been developed which intensified and extended production. These included techniques of irrigation and terracing, particularly common in the Andean area, the use of fertilizers, and the construction of raised fields and floating gardens (see Ch. 3). By either intensifying or extending cultivation, these techniques enabled larger populations to be supported. The chiefdoms and states that they exploited were depleted. They formed egalitarian social groups

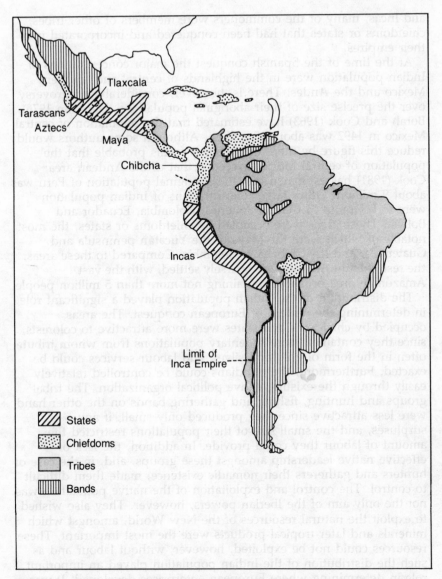

Fig. 2.1 Indian cultures on the eve of Iberian conquest

emerged had populations exceeding tens of thousands distributed in hierarchies of settlements ranging from large urban centres to small villages; Tenochtitlán, the capital of the Aztec empire, is said to have possessed up to 100,000 inhabitants. These societies were politically organized and socially stratified, with a small number of leaders able to command tribute and labour services from commoners. In the case of the major empires, such as those of the Aztecs, Tarascans and Incas, many of the commoners were members of other tribes, chiefdoms or states that had been conquered and incorporated into their empires.

At the time of the Spanish conquest the major concentrations of Indian population were in the highlands of central and southern Mexico and the Andes. There is, however, considerable controversy over the precise size of their aboriginal populations (Denevan 1976). Borah and Cook (1963) have estimated that the population of central Mexico in 1492 was about 25 million. Although some authors would reduce this figure by more than half, it seems probable that the population of central Mexico exceeded that of the Andean area; Cook (1981) has estimated that the aboriginal population of Peru was about 9 million. Other large concentrations of Indian population were to be found in Central America, Colombia, Ecuador and Bolivia. These areas were occupied by chiefdoms or states, the most notable of which were the Maya of the Yucatán peninsula and Guatemala, and the Chibcha of Colombia. Compared to these areas, the rest of Latin America was sparsely settled, with the vast Amazonian area probably containing not more than 5 million people.

The distribution of the Indian population played a significant role in determining the pattern of European conquest. The areas occupied by chiefdoms and states were more attractive to colonists, since they contained large sedentary populations from whom tribute, often in the form of gold and silver, and labour services could be exacted. Furthermore, these Indians could be controlled relatively easily through the existing native political organization. The tribal groups and hunting, fishing and gathering bands on the other hand were less attractive since they produced only small, if any, surpluses, and the small size of their populations restricted the amount of labour they could provide. In addition, the lack of effective native leadership amongst these groups, and, in the case of hunters and gatherers their nomadic existence, made them difficult to control. The control and exploitation of the native population was not the only aim of the Iberian powers, however. They also wished to exploit the natural resources of the New World, amongst which minerals and later tropical products were the most important. These resources could not be exploited, however, without labour and as such the distribution of the Indian population played an important role in determining where European enterprises developed. It was not the only factor that influenced the distribution of economic activities, however, for minerals had to be mined where they were located, and if they were found in areas of sparse Indian population, such as in northern Mexico and Brazil, labour had to be imported from other regions or from Africa.

The colonial period

Colonial policies and practices

The conquest and colonization of Latin America must be seen in the context of political developments in the Iberian peninsula. During the late fifteenth century Spain began to rival Portugal in overseas exploration, and the unification of the Crowns of Aragon and Castile provided an added stimulus for the final conquest of southern Spain from the Moors which was completed in 1492. The same year Columbus discovered America, and in many respects the conquest and colonization of the New World can be regarded as an extension of Spanish activities in the Peninsula; Spain with a centralized government, full of missionary zeal, and with the experience of the *Reconquista* was ready to expand overseas. Similarly, the colonizing activities of the Portuguese in Madeira and West Africa, provided them with valuable experience for the conquest and colonization of more distant colonies. Although Portugal gained a foothold on the South American mainland as a result of Cabral's expedition in 1500, she was more interested in developing trading links with the East, and as such showed little interest in developing Brazil until gold was discovered there at the end of the seventeenth century. Lacking the necessary financial and human resources, the Spanish and Portuguese Crowns made contracts with conquistadors for the conquest and colonization of the New World. Conquistadors provided the financial backing for expeditions in return for which they received from the Crown certain privileges, such as the faculty to found towns and make land grants. In this way the state and private enterprise co-operated to their mutual advantages in undertaking an otherwise impossible task. Nevertheless, in this co-operation the state was the dominant partner, and after the New World had been effectively conquered, both Iberian powers began to restrict the privileges of early conquistadors and colonists, establishing effective administrative bureaucracies and laws that aimed at controlling every aspect of colonial life. The administration in Spanish America was far more extensive than in Brazil. The region was divided into two viceroyalties – the Viceroyalty of New Spain and the Viceroyalty of Peru – with the boundary running across the Panamanian isthmus, and these in turn were divided into *Audiencias* presided over by presidents. Within the jurisdiction of the *Audiencias* governors, *corregidores* and *alcaldes mayores* were appointed as the most important administrators, together with a host of legal and financial officials. Many of these officials had purposely overlapping jurisdictions which effectively established a form of internal spy system which kept a check on their conduct and promoted the power of the Crown (Haring 1963).

However, the introduction of colonial rule in Latin America not only involved the establishment of an administration, as it did in the

case of the British rule in India, but it also involved colonization.
The colonization of Latin America, however, differed from that in
other settler colonies, such as North America, Australia and New
Zealand, in that almost all of the territory was occupied, often by
large native populations. Colonization in Latin America was
therefore more complex. Although colonial activities were controlled
by law, in practice there were considerable regional variations in the
form that it took, which to a large extent reflected the sub-
continent's pre-Columbian heritage. The prime aims of the Spanish
and Portuguese Crowns in America were to civilize and convert the
native population, as well as exploit it and the resources of the New
World for their own benefit and that of their subjects. Conquistadors
and colonists were more exclusively interested in seeking fame and
fortune to enhance their financial and social positions. Initially, their
attention focussed on areas which were inhabited by dense Indian
populations or which possessed easily worked deposits of gold and
silver. Once these quick and easy sources of profit had been
exhausted, colonists turned their attention to economic activities
which required greater investments of time and capital, and which
resulted in more permanent settlements. Silver mining became the
produit moteur of the colonial economy, whilst agriculture developed
to meet European demands for tropical crops and to supply the
growing domestic markets in the towns and mining areas.
Colonization was concentrated in these areas, whilst other areas
which did not produce items for which there was a European or
local demand attracted few colonists. Notable amongst these were
large parts of Uruguay, Argentina and southern Brazil, which
produced essentially temperate products for which a European
demand did not develop until the Industrial Revolution. As such,
these regions attracted few European settlers, the most conspicuous
arrivals being missionaries who were entrusted with the task of
converting and civilizing the small Indian populations that inhabited
these areas, and cattle, which multiplied rapidly on the grasslands.
Thus in the colonial period, it is possible to divide Latin America
into core areas, where European enterprises and the Indian
population were concentrated, and peripheries which remained
largely uncolonized; the location of these cores and peripheries
altered in the Independence period as the demand for products in
Europe changed.

Core areas

Urban centres
Towns and cities were primary instruments of colonization
particularly in Spanish America where they were centres of
political and economic power; in Brazil this role was assumed by the
large estates or plantations, at least up until the second half of the
eighteenth century when the Portuguese Crown tried to enhance the

prestige of the cities in an attempt to counteract the power of rural landlords. As in southern Spain during the Reconquest, towns and cities in Spanish America were employed as symbols of territorial possession and centres from which the surrounding countryside could be administered and exploited. The lowest level of administration was the town council, known in Spanish America as the *cabildo* and in Brazil as the *câmara*, which was responsible for the allocation of land grants, setting market prices etc.; there was no form of rural government for the jurisdiction of one town extended to that of another. Spaniards were required to live in urban centres in the belief that an urban way of life was more civilized and that the residential segregation of Spaniards and Indians would protect the Indians from exploitation and discourage the emergence of powerful feudal estates. The distinction between the urban Spaniard and the rural Indian was never fully realized, however, since in many cases Spaniards did reside on their estates, whilst Indians increasingly migrated to the towns in search of employment. Nevertheless, it is clear that initially power and wealth was concentrated in the hands of Spaniards in the towns, such that Morse has characterized urbanism during the colonial period as centrifugal where power and wealth passed from the urban areas to the countryside; he contrasts this with urbanism in the post-Independence period which he describes as centripetal, with power and wealth passing back to the urban areas with industrialization.

Apart from the administrative functions performed by all towns, some towns performed special roles, but the most important towns were those that were political and cultural centres. Few towns, with the exception of the mining towns and the major ports, played economic roles; few would have survived as trading centres independent of the demands that the Spaniards made for minerals and a limited number of tropical products. Mexico City and Lima, as the capitals of the two viceroyalties, with the seats of archbishops, universities, convents and hospitals remained the most important cities in Latin America throughout the colonial period. Their populations were only surpassed for a brief period by the silver mining town of Potosí, which at the height of silver production in the early seventeenth century was the largest city in Latin America with an estimated population of over 160,000 inhabitants. Potosí was one of the few towns in Spanish America that performed an essentially economic role; the majority of towns were political entities that were imposed on an underdeveloped economic base. Where the local economy developed to meet the demands of the European, and to a lesser extent domestic market, trade stimulated their growth, but they were always vulnerable to changes in market demand and trading policies, as was clear at the end of the eighteenth century when the Spanish monopolistic trading system was reformed. Up until that time, the ports of Veracruz, Portobello, Cartagena and Acapulco flourished as the only ports at which

Spanish fleets could call, whilst the development of Buenos Aires as a port was stifled by a ban on trading there. With the liberalization of trade in 1778, the privileged position of the former ports was destroyed and they went into decline, whilst Buenos Aires entered a period of prosperity. Similarly, the ban on inter-regional trade, by reducing economic competition between towns encouraged stability in the urban settlement pattern, but it prevented the development of regional economies with well-developed urban hierarchies. With the introduction of *comercio libre* in 1778, the protection from competition was removed and many towns went into decline. Significantly, the major towns in Spanish America today remain those which were major administrative centres – the seats of the viceroyalties, *audiencias* and *gobernaciones* – rather than those that performed an economic role.

The dominance of political motives in the foundation of towns and cities was also reflected in their location and form. The first towns to be established were in the Caribbean – Santo Domingo was founded in 1496, Havana in 1514 and Panama City in 1517 – and it was from these centres that expeditions to the mainland were mounted. On the mainland, towns were generally founded in areas of dense Indian population, that is generally in the highlands, and many were located on or near Indian settlements. Mexico City was founded on the site of the Aztec capital of Tenochtitlán in 1521, and Cuzco, the Inca capital, was converted into an effective Spanish town in 1534. Bogotá, Quito and Santiago de Guatemala were all located in areas of dense Indian population.

The Spanish not only established towns in the interior highlands, but they also founded ports on the coast, which in the early years of conquest acted as supply centres through which the Spanish in the highlands were maintained. Later they operated as ports of export for minerals and tropical crops and centres through which European manufactures were imported. The major ports of Veracruz, Portobello and Cartagena have already been mentioned as ports of call for the Spanish fleets, but other notable foundations on the Pacific coast were Guayaquil, Lima/Callao and Arica, the latter being the port through which the Potosí silver was exported. The location of Brazilian towns differed from those in Spanish America in that the majority were located on the coast. This pattern reflected that which existed in Portugal and the fact that the Indian population of Brazil was relatively small and therefore an insignificant factor in determining the location of towns. Only with the discovery of gold in the Minas Gerais in the early 1690s were towns established in the interior.

Consistent with its desire to control every aspect of colonial life, the Spanish Crown insisted that towns established in the New World were to follow a specified pattern – the grid plan. Although early conquistadors had been ordered to establish towns following the grid plan, it was not until 1573 that its precise form and dimensions were specified. The centre of the town was to be marked

Fig. 2.2 Eighteenth-century plan of Quito, redrawn from an original map in the Archivo General de Indias, Seville. 1. Plaza mayor; 2. Cathedral; 3. Jesuit church, La Compañía; 4. Franciscan monastery; 5. Dominican monastery

by a *plaza* around which public buildings, such as the governor's palace, cathedral and town hall, were to be located. These were surrounded by rectangular blocks, the most central of which were generally occupied by the town's most prominent citizens, whilst the lower classes lived on the outskirts of the town, where separate Indian suburbs were also established. Although the form of towns and cities varied in detail, the grid plan was followed with such regularity that any visitor to Spanish America today will be impressed by the monotony of the structure of urban settlements at all levels of the urban hierarchy, from the smallest provincial town to the capital city. Exceptions to this rule were the mining towns, such as Potosí, Tegucigalpa and Taxco, which grew rapidly as irregular settlements before a town plan was drawn up and they were formally established as towns and cities with the titles of *villa* or *ciudad*. Their irregular form was also encouraged by the mountainous terrain in which many of the mining settlements were located. These irregularly developed towns are a welcome deviation from the pattern generally found in Spanish America, and they are more like those found in Brazil. Brazilian towns followed no regular plan but settlements grew up as clusters of dwellings around the major public buildings or a fort; defensive considerations were more important in the siting of towns in Brazil than Spanish America, primarily because of the threat from the Dutch and French to the north. The lack of a formal plan also reflected the less pervasive control of the Portuguese Crown over colonial affairs and the smaller emphasis it placed on urban life and on the town as an instrument of colonization.

The colonial economy

The early sixteenth century

Following the establishment of towns, *encomiendas* and land grants were allocated to those who had taken part in conquest. An *encomienda* was a grant of Indians to an individual who, in return for providing the Indians with Christian instruction, could levy tribute from them in the form of money or goods, and until 1549 could also demand labour services from them. Although the grant of an *encomienda* did not carry with it the land on which the Indians lived, the allocation of *encomiendas* met colonists' demands for immediate wealth and in part satisfied their desire for the type of overlordship that characterized feudal estates in Spain. Although land grants were made at the time of conquest, agriculture was slow to develop given that there were more immediate sources of wealth to be found in the Indian treasures, particularly of Mexico, Colombia and Peru, and the panning of alluvial gold in the Greater Antilles and Central America; agriculture required an investment of money and time, neither of which conquistadors could afford if they were to return to Spain with improved economic and social positions within their short lifetimes. The early sixteenth century can therefore be characterized as one of plunder, where Indian goods were expropriated and Indian labour was often worked to extinction. Once these sources of wealth had been effectively exhausted, the Spanish were forced to turn to less immediate, but often profitable, sources of wealth in the development of silver mining and agriculture.

Mining

Once the Indian treasures had been plundered and the rich and easily worked alluvial gold deposits had been exhausted, the Spanish were forced to turn their attention to the mining of vein ores. Whereas alluvial gold mining required little capital outlay, particularly since Indian slave labour was generally used to pan gold, the mining of vein ores required considerable capital investment. Capital was needed for the construction and maintenance of mining galleries and ventilation shafts, and for the

Plate 1   Colonial mining settlement in south-west Bolivia (1967). San Antonio de Lípez was abandoned late in the eighteenth century as mining became costly and ore more difficult to extract. This has left the settlement form of the seventeenth century intact. (Photo: D. A. Preston)

purchase of equipment and materials to mine and process the ores, as well as for the wages and food of mine workers. The greater capital outlay of vein mining and the more extensive nature of deposits that were mined meant that mining camps became permanent features of the colonial landscape, some of them developing into prosperous towns.

Mining, like conquest, was a joint enterprise between the Crown and private individuals. Whilst the Crown owned all of the sub-soil, it permitted individuals to mine ores in return for one-fifth or a *quinto* of the gold or silver produced. In order to control mining activities, all claims to mines had to be registered with the authorities and all miners swore to bring any bullion they produced to the assay office for taxing. One way in which the authorities attempted to ensure that the Crown's *quinto* was paid was by controlling the allocation of mercury. Initially silver and gold was smelted in crude furnaces, but only rich ores could be processed by this means. In 1554 the amalgamation process was discovered which enabled poorer ores to be refined. This process required the application of mercury and salt among other things, and from the mid-sixteenth century most ores mined in Spanish America were refined by this process. Initially the mercury was imported from

Almadén in Spain, but after the discovery of the Huancavelica mercury mine in what is now central Peru in 1574, over which the Crown established a royal monopoly, most of the mercury came from there. Since most ores could not be worked without the application of mercury, by monopolizing its production and sale, the Crown could keep an account of the amount of silver and gold that was produced and ensure that it received its due fifth.

Unlike agriculture, the location of mining activities was fixed by the distribution of minerals and was less dependent on the availability of local labour, shortages of which could be overcome in many cases by importing labour from other regions. Clearly mine owners sought the cheapest source of labour available, which was forced Indian labour. The Potosi mines were located near to areas with dense Indian populations and the Spanish developed the *mita*, a system of forced labour instituted by the Incas, to supply the mines with labour. Under the *mita* each Indian village was required to supply a quota of its tributary population to work for fixed periods of time for fixed, but low, wages; at the height of production the Potosí mines employed 13,000 *mitayos*. But not all mines were located where there were dense Indian populations. Mining was sufficiently profitable, however, to support the employment of more expensive free labour, as was the case in northern Mexico, or the importation of Negro slaves as occurred in the Brazilian and Colombian gold mines. In these areas, therefore, mining extended European colonization into remote and sparsely settled regions, where it changed the racial composition of the population.

Mineral production in the sixteenth century was dominated by the silver mines of Potosí in Upper Peru and Zacatecas and Guanajuato in northern Mexico (Brading and Cross 1972). From 1531 to 1600 over 15 million kilos of silver were exported to Spain and in the last quarter of the sixteenth century silver bullion accounted for about 90 per cent of Latin American exports by value, with over three-fifths of the silver coming from the Peruvian mines. During the seventeenth century mineral production fell with the decline being more marked in Peru, such that when production revived in the eighteenth century three-fifths of the silver was being produced in Mexico, and in 1800 it minted five times as much silver as had been produced in Peru in 1632.

Mining activities had a number of important political and economic consequences. The profits from mining stimulated the price revolution in Europe and enabled the Spanish Crown to secure loans which were partly used to finance costly foreign wars. From the early seventeenth century Crown revenue increasingly failed to cover expenditure, but its income from the taxes on precious metals effectively enabled it to maintain its empire despite its increasing bankruptcy. On the other side of the Atlantic the *quintos* represented a permanent loss of wealth, but mining often had the effect of stimulating the local economy. The mining industry required mules

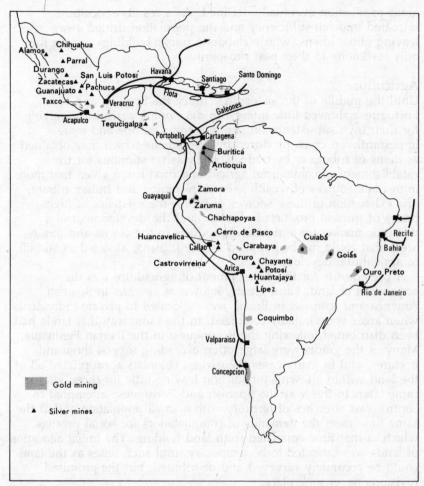

Fig. 2.3 Colonial mining areas and trade routes

for transport, hides for a variety of types of equipment, tallow for candles, as well as food and clothing for mine workers. As a result mining stimulated agricultural development in surrounding or nearby regions, most notably in northern Mexico, in the Bajío of Guanajuato and the valley of Guadalajara, and in north-west Argentina, where annual livestock fairs were held from which the mines of Potosí were supplied. Mining also underlay the prosperity of many towns. Their wealth was reflected in their elaborate architecture and developments in the arts, and in the presence of affluent miners and merchants who formed a flourishing market for luxury goods, including European manufactures and locally produced wine and brandy. The movement of minerals, of goods required in the mining industry, as well as luxury items formed the basis of regional trade. Nevertheless, even in the richest mining areas production eventually declined. As a result, agriculture retreated into self-sufficiency and the population drifted away leaving ghost towns, whose elaborate public buildings remain the only testimony to their past prosperity.

Agriculture

Until the middle of the sixteenth century the Spanish and Portuguese showed little interest in developing agriculture, relying for their own subsistence on supplies from Europe and more importantly on crops produced by the Indians which they obtained as items of tribute or by trade. There was no stimulus for the establishment of commercial agricultural production given that more immediate sources of wealth existed in mining and Indian tribute. The exhaustion of these sources of profit, the existence of high prices of tropical products in Europe and the development of a domestic market for agricultural products in the towns and mining areas that could not be supplied by the Indians, all acted as stimuli to agricultural development.

A prerequisite for the development of agriculture was the acquisition of land. Land grants, known as *mercedes* in Spanish America and *sesmarias* in Brazil, were allocated to private individuals when areas were initially colonized, in the same way that lands had been distributed following the Reconquest in the Iberian Peninsula. Many of the grants were large often exceeding several thousand hectares, and in many cases the earliest colonists appropriated all of the land within a town's jurisdiction leaving little for those who came later. In this way the Spanish and Portuguese attempted to control vast stretches of territory with a small population and at the same time meet the demands of conquistadors for social prestige, which at that time emanated from land holding. The initial allocation of lands was intended to be temporary until such times as the land could be accurately surveyed and distributed, but the promised revisions never took place.

In the allocation of lands, the rights of Indians to their lands was protected in law. However, the law only recognized private ownership of land, and customary rights, such as those exercised by tribal groups, were not recognized. Whilst Indian lands were often usurped, in other cases Indians sold their lands in order to discharge tribute debts or meet other official and non-official demands of the state and church.

Although the initial land grants were large, many land holdings were expanded piecemeal. Particularly active in this field were the missionary orders, notably the Jesuits and Dominicans, whose land holdings increased as those owned by secular landlords were reduced in size by inheritance laws which demanded their subdivision amongst heirs. Increasingly popular, however, was the process of entailment by which a landlord pledged an annuity to a church or a convent and thereby ensured that the land would be passed on as a unit and not subdivided. In some cases landowners expanded their holdings with the aim of monopolizing land and driving out competition, and, particularly where labour was short, creating a landless labour force driven to depend on wage labour. Once in their employment, landlords attempted to maintain workers in their employment by encouraging them to incur debts, a system known as debt peonage, or by giving them plots of land on their estates. The extent to which Indians were tied to estates by debts or other means has generally been exaggerated and probably a more powerful force retaining Indians on estates and in developing the mutual dependence between landlord and worker, was the employees' lack of alternative means of subsistence. Although this pattern of employment was common in the colonial period, it probably increased during the nineteenth century. Other estate owners employed Indians they held in *encomienda* to work in their fields, even though the practice was illegal after 1549, whilst in regions of dense Indian population, such as Oaxaca, highland Guatemala and highland Ecuador, the *repartimiento*, a system of forced labour similar to that of the *mita* used in mining, continued to operate throughout the colonial period. In areas where there were few Indians, but which were suited for the production of profitable tropical crops, particularly sugar, it was an economic proposition to import expensive Negro slaves. These areas included the Greater Antilles, where the native population had been decimated in the early years of conquest, Brazil and the Guianas. It has been estimated that during the colonial period about 11 million Negro slaves were imported to Latin America, mainly from Angola and the Guinea coast, with the majority introduced illegally through contraband trade. All of the areas into which slaves were introduced in large numbers were located near to the sources of slaves in Africa and near to European markets; although other areas, such as coastal Peru and Ecuador, were equally suited to the cultivation of tropical crops, high transport costs reduced the profitability of their products, such that the importation of Negro slaves on a large scale was not a viable proposition for most *hacendados*.

Before presenting a brief survey of the spatial distribution of commercial agricultural production, it is worth examining a theoretical question that has preoccupied the literature in recent years (Mörner 1973; Van Young 1983). Was the hacienda a capitalist or feudal institution? Some authors, notably André Gunder Frank (1979), maintain that the hacienda was capitalistic from its inception, arguing that it was market-oriented, and not self-sufficient as a number of other authors have suggested. Whilst Frank is correct in asserting that haciendas did produce for a market, whether European or domestic, market-orientation is an inadequate basis on which to classify them as capitalistic, for feudal estates often produce for a market, albeit a local and limited one. Any characterization of the hacienda as a capitalistic institution must take account of the existence or not of free wage labour. Whilst free wage labour began to emerge in Latin America during the colonial period, it certainly did not exist in the sixteenth century, and even when it began to emerge its full development was retarded in a number of ways. First, royal legislation restricted the monopolistic control of land by protecting Indian rights to their lands, albeit imperfectly, and second, Indians were often given land on which to raise crops for their own subsistence, so that they were not wholly dependent on wages and therefore free to sell their labour power. Furthermore, to suggest that the hacienda was capitalistic from its inception, implies that capitalism developed earlier in Latin America than in Europe. Although the hacienda cannot be characterized as wholly capitalistic, neither can it be described as entirely feudal. If the estates had been feudal they would have been largely self-sufficient and dependent on the exaction by a powerful landlord of rent, in labour or production, from serfs who were owners of their land. As has already been indicated, haciendas were not generally self-sufficient, but they did produce for a market, and as such landlords acquired not only prestige and power from the mere acquisition of land, but often reasonable incomes. The fact that many haciendas were only farmed in part was probably due as much to the desire of estate owners to limit production and maintain high prices, as it was to their lack of interest in land other than the prestige it bestowed. It is perhaps more accurate, if somewhat unsatisfying, to conclude that the hacienda possessed both feudal and capitalistic elements the balance of which varied from region to region.

The nature and distribution of commercial agricultural production was strongly influenced by the demand for particular products and by the ability of areas to produce them. Many of the crops and most of the animals that were raised were not indigenous to the New World. Whilst cacao, tobacco, and dyestuffs, such as indigo and cochineal, had been produced by the Indians in pre-Columbian times, sugar, the citrus fruits, wheat and barley were introduced from Europe in the sixteenth century; coffee and Asian spices, such as cinnamon, did not arrive until the eighteenth century, when they were introduced from the East Indies by the Dutch. Although these crops were rarely adopted by the Indians, some of them became important export crops. Since the Indians had possessed few domesticated animals in the pre-Columbian period, cattle, sheep, horses, pigs and chickens contributed greatly to agricultural production both for export and subsistence, with cattle in particular multiplying rapidly on the open grasslands. The demand for agricultural products in Europe consisted mainly of tropical crops such as sugar, cacao and dyestuffs, as well as hides, whilst the expanding urban populations in Latin America constituted a domestic market for food, particularly wheat and maize, and the mining industries required hides, mules and tallow. The lack of demand for temperate products at this time meant that agricultural production in Uruguay, Argentina and Chile was essentially oriented towards the domestic market, which in those areas remained small. Demand and the physical ability of areas to produce particular crops,

however, were not the only factors involved. The production of
certain products, such as wool, oil and wine, for which there was a
demand in Europe, was specially discouraged in Latin America
because production there would have competed with that in Europe.
In addition, the commercial production of tropical crops in the more
distant Latin American provinces was prohibited by high transport
costs. Hence, sugar production in coastal Peru was oriented towards
the domestic market, because it could not compete in European
markets with that cultivated in the Caribbean.

Problems of transportation generally meant that agricultural
production in mainland South America focussed on the domestic
market which was located in the major towns and ports, as well as
the mining areas. The major market in Peru was the mining centre
of Potosí, which consumed food, wine and brandy produced in the
coastal oases and central Chile, and was supplied with livestock and
animal products from north-west Argentina. In other areas estates
produced wheat, maize, and livestock, to support the towns, ports
and lesser mining centres; the high Andean basins of Peru supplied
the mines of Cerro de Pasco and Huancavelica, as well as Lima,
whilst the highland basins around Quito and Bogotá supplied the
local mining industries of Zaruma and the Upper Cauca valley
respectively. The Mexican economy was more diversified. Mexico
exported products to Europe and supplied the domestic market in
the towns and mining areas. The production of food first
developed to supply the expanding urban population of central
Mexico that had formerly been supported by the Indian population,
which had declined. Beginning in the Valley of Mexico agricultural
production spread rapidly into the Bajío and the Valley of
Guadalajara as silver mines were opened up to the north. In the
Caribbean and Brazil, agriculture was dominated by the production
of sugar, although in Brazil livestock raising developed in the
interior; at first it provided mules for the sugar industry, but later,
from the early eighteenth century, it supplied the needs of gold
miners in the Minas Gerais.

Manufacturing industry

Despite the fact that in pre-Columbian times the Indians in many
parts of Latin America had been skilled potters, weavers and
silversmiths, manufacturing industry was slow to develop during the
colonial period. The Spanish Crown aimed at protecting its home
industries by establishing a monopoly of trade in European
commodities. Unfortunately, the home industries were unable to
meet the demand for such items as fine cloths, leather goods, glass,
pottery, weapons and general hardware, so that many goods had to
be imported from other parts of Europe before being shipped across
the Atlantic from Spain. Even so, the supply was insufficient to
meet the demands of colonists, who were often forced to turn to
contraband trade with the English, French and Dutch in order to
obtain the items they required. The development of local industries

that might have met this demand was often discouraged and even suppressed in order to protect home industries and trade, but interference in colonial industries was never as systematic or as complete as it was with trade, and a large number of small-scale enterprises did develop to supply the local market, which included many Indians who were generally unable to purchase expensive European manufactures. Apart from the processing of foods and beverages, and the refining of silver, the most important manufacturing industry to develop in the New World was the textile industry. Coarse woollen and cotton cloth were produced in east-central Mexico supplying the mining areas to the north, but the most renowned area of production was in the highlands of Ecuador around Quito, Latacunga and Riobamba, from which textiles were transported to Peru and the Potosí mining area, where they supplemented those produced locally.

Trade and transport

In the colonial period trade and transport were export-oriented. Apart from the trade across the Atlantic, commercial links within the sub-continent were dominated by the movement of goods between Europe and regions whose economies were geared to the European market. The outward orientation of production altered many of the road systems that had existed in pre-Columbian times. For example, the Spanish developed the west-east Inca roads which focussed on the ports, at the expense of the north-south routes that the Incas had emphasized in integrating their empire. Whilst those roads that could not meet colonial demands rapidly fell out of use, except by the Indians, new roads were built in formerly underdeveloped areas, such as northern Mexico, but only where they could be economically or politically justified. The basis of the colonial transport network was the *camino real* or royal highway; with the exception of the Magdalena river in Colombia, the Spanish did not develop river transport. Goods were transported mainly by mules and ox-drawn carts.

Although these forms of transport represented an impovement over the llama – the only form of transport in pre-Columbian times that was restricted to the Andean area – mules and carts were slow and not suited to bulk transport. Mules only averaged about 19 to 24 km (12 to 15) miles a day and could only carry about 136 kg (300 lb). This form of transportation limited the production of perishable goods and those with a high ratio of weight to value. It also meant that the commercial production of crops for the European market was not a viable proposition in remote parts of South America; even in more accessible regions, such as Mexico, agricultural production favoured those crops of high value and little weight, such as indigo and cochineal. The limitations of the transport system also effectively restricted large scale production of crops for a mass domestic market, so that estates were forced to remain relatively small producers whose products were rarely traded beyond neighbouring provinces.

The movement of goods to and from Spanish America was controlled by the Crown, which established a monopolistic and protectionistic trading system aimed at securing the greatest profits for itself and for the few merchants who it entrusted with its operation; foreigners were not allowed to trade in Spanish America, except that they could introduce Negro slaves. The idea was that Spain would supply the colonies with manufactures, including luxury goods, in return for the minerals and tropical agricultural products they produced. European goods were to be transported on Spanish vessels manned by Spanish sailors, after having been registered in Seville, where the Board of Trade, the *Casa de Contratación* (established in 1503), and the *Consulado*, the merchant guild which arranged for their shipment, were both located. All goods destined for Spanish America from anywhere in Europe had to pass through Seville (from 1717 through Cadiz), and as such merchants controlled all trade with Spanish America and by setting the prices for goods ensured maximum profits for themselves. To safeguard goods in transit, from 1552 ships were required to sail in convoy under the protection of armed vessels known as *armadas*. Two fleets were dispatched from Spain every year. One, the *Flota*, sailed to Veracruz in Mexico, convoying ships for Central America. The other fleet, the *Galeones*, sailed to Portobello and Cartagena. Through Portobello goods passed across the Panamanian isthmus to the Andean and river Plate countries. Up until 1778 no trading could take place through Buenos Aires, which remained a closed port in order to prevent contraband trading with the Portuguese and Northern Europeans. Through Cartagena goods passed to Colombia and Venezuela. At these three ports merchants, mainly from Lima and Mexico City, monopolized the items coming from Europe and from thence transported them to the colonies selling them at highly inflated prices. The two fleets wintered in Latin America picking up minerals and agricultural produce to transport to Spain, after which they rendezvoused in Havana before sailing across the Atlantic together. Just as the trade across the Atlantic was restricted to certain ports and controlled by a handful of merchants, so trade with the Philippines could only take place through Acapulco.

The fleet system operated until the beginning of the eighteenth century, but it reached its peak about 1590, when the fleets comprised thirty to ninety vessels, after which it gradually disintegrated. Spain could not provide the manufactures demanded in the colonies – by the end of the colonial period she imported nine-tenths of the manufactures she exported to Latin America from other countries in Europe – and as such colonists turned to contraband trade. This represented an increase in expenditure to the Crown as well as a loss of revenue. In addition, the provision of *armadas* to protect the fleets was paid for out of increased duties on goods, which made them more expensive than those offered by foreigners. Other goods were lost through vessels being caught in hurricanes or being attacked by pirates. With the disintegration of the fleet system, the Crown established equally unsuccessful privileged trading companies, such as the Caracas Company, and finally in 1778 trading restrictions were lifted, allowing ports in Spain and Latin America to trade freely with each other. At the same time duties on goods were reduced thereby discouraging contraband trade, so that altogether the reforms are said to have resulted in a 700 per cent increase in Crown revenue from trade.

The commercial system developed for Brazil was never as clearly formulated or as extensive as that developed for Spanish America, for Portuguese commercial interests focussed on the East. A fleet system was introduced in the middle of the seventeenth century in an effort to counteract the presence of foreigners, but it was never as restrictive as that developed by Spain, and it underwent many changes before it was finally abolished in 1765.

The peripheries

The discussion hitherto has focussed on areas where the Spanish and Portuguese established their towns and economic enterprises, drawing into their orbit the indigenous population. Not all areas, however, were attractive to Europeans during the colonial period and these areas remained outside effective administration. These were essentially areas where the Indian population was sparse and where the resources were not in demand in Europe at that time. Developments in these areas can be divided into three types: first, the sparsely inhabited grasslands, particularly of Argentina, and southern Brazil and southern Venezuela, were subject to invasion by feral cattle. During the colonial period these cattle were hunted by small numbers of marginal people of mixed race, but in the nineteenth century these areas emerged as the major ranching regions of Latin America; second, in many areas missions were founded to convert and civilize the Indians; third, Northern Europeans occupied territories where Spanish and Portuguese control was weak. The colonization of the latter two areas will be considered in more detail.

The missions
The initial conversion and civilization of Indians living in remote parts of Latin America was undertaken by the missionary orders. The intention was that missionaries would move into an area and establish missions to which they would attract the Indians who they would civilize and instruct in the Catholic faith. Theoretically, after ten years the missions were to be handed over to the secular authorities and the missionaries were to move on to more remote areas civilizing and converting the Indians. Nevertheless, because of the shortage of parish priests and secular officials to administer the mission settlements, they very often remained in the hands of the missionaries for longer periods. The missions acted as effective instruments of colonization, whose activities were supported

financially by the Crown, particularly in areas which were under threat of foreign domination, such as northern Mexico, eastern Central America, the lowlands to the east of the Andes and the La Plata region. The most important orders that established missions in Latin America were the Jesuits, Franciscans, Dominicans and Capuchins. The task of establishing missions was not easy since the Indians were often dispersed over wide areas and their nomadic existence made them reluctant to settle and remain in the missions. As a result soldiers were often required both to establish the missions and to retain the Indians there. Apart from converting the Indians, the missionaries instructed the Indians in agricultural techniques. They also required polygamous groups to adopt monogamy and, by establishing Indian *cabildos*, they introduced them to a form of political government. Thus, missionization resulted in Indian cultures being profoundly modified, if not destroyed. This coupled with the introduction of Old World diseases resulted in the decline and disappearance of many Indian groups. Even in Paraguay, where the Indians fled in large numbers into the missions to escape the enslaving expeditions of the Paulistas, the expulsion of the Jesuits in 1767 resulted in the disintegration of the mission settlements, as deculturated individuals drifted away in search of employment on local estates or in the towns. Hence, although the missionaries considered themselves to be protectors of the Indians, in fact in most cases, they were unconscious forces for their destruction.

The northern Europeans

For both ideological and practical reasons the Iberian powers sought to prohibit the settlement of foreigners in the New World. This policy was consistent with the prevailing concept of absolute monarchy and it was justified by the Papal Bulls of 1493 and the Treaty of Tordesillas of 1494 which upheld Spanish and Portuguese claims to the New World, specifically for the propagation of the faith. The Iberian powers also sought to exclude foreigners to limit the economic and political power they might have acquired given access to the resources of the New World. The northern European nations, notably England, France and Holland, on the other hand, were anxious to undermine Spanish and Portuguese power in Europe, and their American colonies were the most vulnerable parts of their empires. Initially foreign activities comprised attacks on the treasure fleets and the ports, but later from the early seventeenth century interest focussed on acquiring footholds on the Caribbean islands and the fringing mainland from which they attempted to undermine Iberian power by contraband trade and later colonization. These areas, with the Guianas, were those where Spanish and Portuguese control and administration were the weakest, primarily because they attracted few settlers since they possessed few resources and only sparse Indian populations. The English first settled in Barbados in the 1620s, but probably the most important

English possession in the Caribbean was Jamaica, which was seized in 1655. Meanwhile the French had occupied Guadeloupe and Martinique and had begun to colonize western Hispaniola, whilst the Dutch had seized Curaçao in 1634. In addition, all three nations jostled to establish colonies on the weakly defended and uninviting Guiana coast. Although most of the seizures and occupations of northern Europeans in the New World were illegal, their rightful possession to these territories was often confirmed at peace treaties concluding European wars. As a result their ownership often changed hands regardless of political developments in the colonies. The acquired colonies were used as bases for contraband trade with the Spanish colonies and Brazil, whilst the development of sugar and tobacco plantations using imported Negro slaves on the Caribbean islands and in the Guianas constituted valuable sources of income that enhanced the economic power of the northern European nations. Even where these nations did not formally acquire territorial possession, their influence was often strong. Such was the case along the Caribbean coast of Central America where the English established dyewood-cutting enterprises.

Indian societies under colonial rule

Iberian conquest and colonization were disastrous for the Indian population of Latin America. At the end of the colonial period even the most fortunate Indian groups that had come into contact with Europeans were less than half of the size they had been at the time of Conquest and some had become extinct. Although the Indian population declined during the colonial period, the demographic changes experienced by different Indian groups varied. Indians in the Greater Antilles and on the fringing mainland of the Caribbean became virtually extinct within a generation, whilst those in Mexico and Central America suffered a dramatic decline in the early colonial period, but by the late seventeenth century their populations had begun to increase. In the Andean area the decline was not as marked as in Mexico, but there the Indian population did not begin to recover until the middle of the eighteenth century. In other parts of Latin America, most of which were inhabited by tribes and bands, the Indian population continued to decline throughout the colonial period.

A number of factors were responsible for the decline in the Indian population: disease; the systematic killing, ill treatment and overwork of the Indians; miscegenation and the disruption to Indian economies and societies caused by conquest and colonization, including their psychological impact. There is general agreement that disease was a major factor in the decline of the Indian population which possessed no immunity to diseases introduced from the Old World. The most notable killers were smallpox, measles, typhus, plague, influenza, yellow fever and malaria. There are numerous accounts of the populations of villages and whole areas being reduced by one third or one half as a result of epidemics, particularly of smallpox and measles, and the devastating impact of these diseases on previously non-infected populations is corroborated by historically more recent epidemics.

Sixteenth-century observers blamed the rapid decline of the Indian population on the ill treatment and overwork of the Indians by conquistadors and colonists. There is no doubt that in the Caribbean the Black Legend was a reality which contributed significantly to the extinction of formerly dense Indian populations. In 1542, as a result of representations to the Spanish Crown, particularly by the Dominicans, the New Laws were introduced which aimed at improving Indian-European relations and protecting the Indians from exploitation. Although there were numerous infringements of the New Laws, by banning Indian slavery, moderating the use of Indian labour and regulating the amount of tribute that could be levied, they did lead to a general improvement in the treatment of the Indians to the extent that the colonization of the mainland which

Fig. 2.4 Ill-treatment of Indians in the mines by Spanish officials, after the seventeenth-century Indian chronicler Guaman Poma de Ayala

occurred mainly after their introduction, did not result in a repeat of the demographic disaster that had occurred in the Caribbean and to a lesser extent in Middle America.

Another factor which became increasingly important in contributing to the decline in the Indian population and retarding its recovery was miscegenation. The imbalance in the sex ratio of colonists to Latin America in the early years of conquest encouraged racial mixing and the emergence of Mestizos (see Ch. 9). Although the Crown passed laws limiting contact between the races, such contact was inevitable given the refusal of Europeans to undertake manual labour and their reliance on Indian, and to a lesser extent, Negro labour. As such the major centres of employment – the haciendas, mines and towns – emerged as racial melting pots.

Equally important in contributing to the decline in the Indian population, but often understressed in the literature, was the impact of the disruption to Indian economies and societies brought about by conquest and colonization. In many cases, Indian economies were unable to meet the demands made upon them with decreased resources. During conquest Indian lands were over-run and pillaged, whilst others were usurped by Europeans. Even where Indians retained their lands, they were often over-run by straying livestock. Not only was the availability of land reduced, but also, and probably more important, the availability of labour to work it. In Peru the irrigation systems developed by the Incas in the pre-Columbian period could not be maintained given the decline in the Indian population, and as such they fell into disrepair. More generally the decline in the Indian population and the high demands from colonists for Indian labour, reduced labour inputs into subsistence activities resulting in food shortages and even famines. The impact of declining production was mitigated in part, however, by the introduction of domesticated animals from the Old World, particularly the chicken and to a lesser extent cattle. These increased the availability of protein at a time when labour inputs into hunting and fishing declined. The impact of declining agricultural production would not have been so serious if demands made upon it had fallen parallel with the population, but in fact they increased because the Spaniards required the Indians to pay tribute, to contribute to the salaries of priests and to purchase often unwanted goods from royal officials. The inability of Indians to meet the demands made upon them encouraged them to sell their lands and seek employment as wage labourers in the towns or on local haciendas. Here they came into contact with other races and gradually lost their cultural and racial identity.

The social organization of Indian groups was altered directly by the Spanish and indirectly through population decline. Whilst Spaniards usurped the power and status of Indian leaders, Indian societies as a whole experienced social levelling. With the exception of Indian leaders, the Spanish did not recognize native social classes, but rather subjected the Indians to common laws and systems of tribute payment. The social organization of Indian communities was also modified by extensive resettlement schemes which aimed at congregating the Indians in large settlements where they could be more effectively administered and instructed in the Catholic faith. Massive resettlements took place in Mexico at the turn of the sixteenth century, whilst in Peru in the 1570s they involved the movement of about 1.5 million Indians. The new settlements often contained remnants of decimated communities which had few interests in common. This weak social organization was undermined even further by the prolonged absence of individuals who worked under the *repartimiento* and by the permanent abandonment of villages by Indians who fled to escape the burdens of tribute payment and forced labour. The prolonged or permanent absence of individuals led to high instability in marriage, whilst the marriage rules of some Indian groups were completely altered through the church's insistence on monogamy.

The psychological impact of conquest and colonization should not be underestimated, even though it is difficult to quantify. Defeated in conquest and ravaged by disease, the Indians believed that they were being punished by their gods. This, in addition to the real hardships of existence, fostered a lack of will to survive, such that abortions and infanticides increased and families became smaller.

The processes of change just described were experienced by the Indians who at the time of Conquest for the most part formed chiefdoms and states. The tribal groups and the hunting, fishing and gathering bands experienced even more profound changes. As already indicated, the difficulties of controlling these groups and the lack of economic interest they provided, meant that their initial conversion and civilization were entrusted to the missionary orders, and the disastrous impact of missionization on the Indians has already been described. In areas which were inhabited by nomadic hunters and gatherers, the Spanish and Portuguese only attempted to control the Indians by enslaving them where they either harrassed and attacked European settlements, or where the regions were under threat of domination by foreign powers. Indian slavery had been banned in 1542, but it continued in remote parts of the Empire, notably in northern Mexico, southern Chile and Argentina, where the Indians proved exceptionally difficult to control. In Brazil Indian slavery persisted throughout the colonial period. Particularly notorious were the Paulistas (from São Paulo) who conducted expeditions into the interior in order to acquire Indian slaves to sell on the coast. Slavery not only resulted in the break up of Indian communities, but by bringing the slaves into close contact with other races it encouraged the loss of their cultural and racial identity.

Conclusion

The concentration of population and economic activities in the highlands of Middle America and the Andes in pre-Columbian times was to a certain degree emphasized during the colonial period, although not without a decline in the Indian population and a reorientation of production. Areas which had been sparsely settled in pre-Columbian times remained as such throughout the colonial period, with the major exception of areas where precious minerals were found. An important spatial change, however, was the development of coastal settlements which acted as political and economic control centres for the administration and economic exploitation of the sub-continent. These spatial changes reflected the aims of the Iberian powers and the nature and distribution of the Indian groups they encountered. The distribution of population and economic activities was to experience perhaps more sweeping changes in the Independence period, when the temperate regions, which had remained underdeveloped during the pre-Columbian and colonial periods, received an influx of capital, technology and immigrants aimed at developing the production of temperate agricultural products for which a demand had developed in Europe with the Industrial Revolution.

Further reading

Bethell L (ed.) (1985) *The Cambridge History of Latin America*, Vols 1 and 2. CUP, Cambridge. (Essential reading on the pre-Columbian and colonial periods. Collections of excellent and up-to-date essays by the most eminent scholars in their fields.)

Bolton H E (1917) 'The mission as a frontier institution in the Spanish American colonies', *American Historical Review* **23**, 42–61. (The classic paper on the subject.)

Borah W and **Cook S F** (1963) 'The aboriginal population of Central Mexico on the eve of Spanish conquest', *Ibero-Americana* **45**. University of California, Berkeley and Los Angeles.

Brading D A and **Cross H E** (1972) 'Colonial silver mining: Mexico and Peru', *Hispanic American Historical Review* **52**, 545–79.

Bray W, Swanson E H and **Farrington I S** (1975) *The New World*. Elsevier-Phaidon, Oxford. (A well illustrated introduction to the archaeology and cultures of pre-Columbian Latin America.)

Cook N D (1981) *Demographic Collapse: Indian Peru, 1520–1620*. CUP, Cambridge.

Denevan W M (ed.) (1976) *The Native Population of the Americas in 1492*. University of Wisconsin, Madison. (A collection of recent papers on the subject including a new hemisphere estimate.)

Frank A G (1979) *Mexican Agriculture, 1521–1630: Transformation of the Mode of Production*. CUP, Cambridge.

Gibson C (1966) *Spain in America*. Harper & Row, New York. (Probably the best introduction to colonial Spanish America.)

Haring C H (1963) *The Spanish Empire in America*. Harbinger Books, New York.

Harris M (1964) *Patterns of Race in the Americas*. Walker and Co., New York.

Hennessey A (1978) *The Frontier in Latin American History*. Arnold, London.

Houston J (1968) 'The foundation of colonial towns in Hispanic America', Ch. 15, pp. 352–90 in **Beckinsale R P** and **Houston J** (eds), *Urbanisation and its Problems*. Blackwell, Oxford.

Lang J (1975) *Conquest and Commerce: Spain and England in the Americas*. Academic, New York. (Compares the character of conquest and colonization in Latin America and North America.)

Mörner M (1967) *Race Mixture in the History of Latin America*. Little and Brown, Boston.

Mörner M (1973) 'The Spanish American hacienda: a survey of recent research and debate', *Hispanic American Historical Review* **53**, 183–216.

Prado C Jr (1967) *The Colonial Background of Modern Brazil*. University of California, Berkeley and Los Angeles.

Sánchez-Albornoz N (1974) *Population of Latin America: A History*. University of California, Berkeley and Los Angeles. (Broader text than the title suggests. About half of the book considers the pre-Columbian and colonial periods.)

Sanders W T and **Marino J** (1970) *New World Prehistory*. Prentice-Hall, Englewood Cliffs, N.J. (A comprehensive introduction to the pre-Columbian archaeology of Latin America.)

Steward J H and **Faron L C** (1959) *Native Peoples of South America*. MacGraw-Hill, New York, London and Toronto. (Although over twenty years old, still the best general introduction to Indian cultures in South America, past and present.)

Van Young E (1983) 'Mexican rural history since Chevalier: the historiography of the colonial hacienda', *Latin American Research Review* **18**, 5–61.

West R C and **Augelli J P** (1976) (2nd edn) *Middle America: its Lands and its Peoples*. Prentice-Hall, Englewood Cliffs, N.J. (Contains excellent sections on the colonial geography of Middle America.)

CHAPTER 3

Indigenous agriculture in the Americas: origins, techniques and contemporary relevance

Richard Smith

At the time of European contact, agriculture was widely practised throughout the Americas from the lands of the northern Algonkians to the temperate regions of Argentina and Chile. It is, however, to the tropical and sub-tropical areas of Central and South America that attention will be focussed, for it is from these physically diverse parts of the New World that much of the evidence for indigenous, pre-Hispanic agriculture has been derived. Not only was the range of crops and agricultural methods greater in this region but we can be fairly certain that it was from this area, or from sub-centres within it, that agricultural impulses spread to areas north and south.

The tropical regions now provide some of the earliest evidence of plant domestication and crop husbandry and in parts of the American tropics prior to the arrival of the conquistadors, indigenous agriculture had reached its climax. In other areas it had disappeared or degenerated centuries before, as in the Maya lowlands, while the development of state and confederate structures in the century or so prior to Spanish contact had led to a mixture of abandonment and reclamation of former agricultural lands. Yet, to use European terminology, American civilizations, culminating in the Maya, Aztec (Mexica) and Inca empires were technologically in the Bronze Age. Because of acknowledged parallels between prehistoric cultures, the study of pre-Hispanic agriculture and society gives us unique insight into the nature of early societies in general. In Europe, several thousand years of landscape change and cultural evolution separates us from the action while in Latin America we are, thanks to sixteenth-century chroniclers, at the threshold of the culture itself (Fig. 3.1).

Pre-Hispanic and contemporary indigenous agriculture reveals not only a range of technical achievement, but also an evident understanding of ecological principles, contrasting with much of what has happened in the last four hundred years. During and after the conquest vast areas of former agricultural land, much of it in geographically remote and less rewarding situations, were

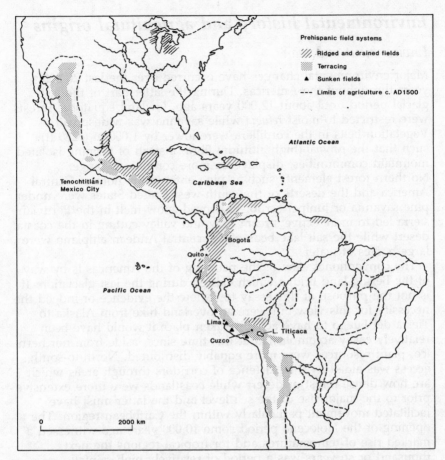

Fig. 3.1 Indigenous American agriculture: some locations and limits (based partly on Donkin 1979)

abandoned, while better quality irrigable lowlands were acquired by large estates. The remainder, representing varying degrees of difficulty for mechanized agriculture, continues even now in peasant holdings in a manner which indicates that some of the past is alive in the present but struggles to survive.

To provide a backcloth for discussing pre-Hispanic agriculture a review will first be given of environmental history which will be followed by an introduction to pre-Hispanic crop plants and to the origins of agriculture. A major section is then devoted to a variety of pre-Hispanic agricultural forms appropriate to different terrain and their significance for local subsistence and wider exchanges. The final sections are concerned with changes in agricultural outlook since the Spanish conquest, problems which lie ahead, and whether indigenous practices have a future role to play.

Environmental history and agricultural origins

Environment and man

Major environmental changes have occurred over the last 20,000 years throughout the Americas. During the latter part of the last glacial period until about 12,000 years ago, lowland equatorial forests were restricted to moist *refugia* while savanna was widespread. Vegetation belts in the cordillera were lower by 1,000 m (3280 ft) such that the *páramo* (high altitude) flora, instead of forming isolated mountain communities, displayed a more continuous range. Northern forest elements such as pine entered the flora of Central America and the deserts of the south-west United States were under pine savanna or juniper scrub. Enhanced snow melt in the Peruvian sierra led to more active stream flow and valley cutting in the coastal desert while the salt lake beds of the central Andean altiplano were large bodies of water.

The conventional view of the peopling of the Americas is by way of the Bering land bridge which existed during the last glaciation. It is not the purpose of this essay to review the evidence or indeed the necessity for this view but were an overland hike from Alaska to Tierra del Fuego to have actually taken place it would have been relatively easily accomplished at this time since, aside from northern ice, plant resources were more equably distributed. North-to-south access was aided by the existence of corridors through areas which are now dense forest or desert while coastlands were more extensive prior to the major rise of the sea level and the latter must have facilitated movement particularly within the Caribbean region. The opening of the Holocene period some 10,000 years ago witnessed a marked rise of temperatures and for tropical regions the next thousand or so years was a period of relatively high rainfall. Temperatures continued to rise until about 6,000 years ago by which time forests had reached their maximum extent and elevation. Evidence of periodic drying of the climate since 8,000 years ago suggests that eustatic change, occurring until about 5,000 years ago, probably provided valuable hydrological compensation in low-lying areas. In any event, American low latitudes were not as conspicuously affected by desiccation as were other tropical areas though reciprocal fluctuations of savanna and rain forest have been detected.

Moist tropical lowlands and arid areas both pose particular problems for palaeoecological work while the paucity of studies in these environments helps to emphasize that statements about man's impact in prehistoric times are still largely tentative. In time, man would have occupied an increasingly warm and forested landscape yet, owing to the ecological importance of the forest-grassland fringe, supposed evidence for climatic drying may increasingly be equated with human disturbance. Sudden later advances of savanna in South America do strongly implicate man while recent studies in the Yucatán improve our picture of the Maya area. Here the earliest Holocene landscapes comprised pine forests and savanna, with man present somewhat before 10,000 years ago possessing a tool kit similar to those found in Panama and Ecuador. This early vegetation was replaced later by broad-leaved forest while about 4,000 years ago savanna expanded once again. Particularly marked was the reduction of tree vegetation and rise of agricultural indicators prior to the Maya Classic period. Thick clay layers are widespread and associated mainly with the latter whilst Maya collapse in the ninth century prompted a rapid and widespread advance of forest within about 150 years.

Plant domesticates

The diversity of plants utilized on Spanish contact is impressive and a brief cross-section is offered here. There were plants such as totora reeds (*Scirpus* sp.) and cat's-tails (*Typha* sp.) whose fibre provided raw material for the construction of boats and dwellings. The rhizomes of these species were also a potential source of food. Cotton (*Gossypium hirsutum* and *G. barbadense*) was the basis of textiles and items such as fishing nets. There were roots and tubers of forest origin such as manioc (*Manihot esculenta*) and sweet potato (*Ipomoea batatas*) together with yautia, the American variant of taro (*Xanthosoma* sp.) and yam (*Dioscorea trifida*). The Andes were the home of the potato (*Solanum tuberosum* and other species) while achira (*Canna edulis*) was one of many roots grown locally but of less widespread importance than the sweet potato. The ubiquitous cucurbits were represented by the common gourd (*Cucurbita* sp.), squash (*C. moschata*) and the bottle gourd (*Lagenaria siceraria*). The latter was used for utensils while gourds and squash were used as moulds for pottery and probably as floats in fishing. The legumes were represented by common beans (*Phaseolus vulgaris*), lima beans (*P. lunatus*),jack beans (*Canavalia* sp.) and the peanut (*Arachis hypogaea*). Grain crops included maize (*Zea mays*), probably the most famous of all American cultigens, but also amaranths (*Amaranthus* sp.) and the chenopods, quinoa (*Chenopodium quinoa*) and cañihua (*C. pallidicaule*). Certain species were prized as spices and flavourings,for example chilli peppers (*Capsicum annuum*) and vanilla (*Vanilla planifolia*), the climber of Mesoamerican forests. Some were of medicinal use, including the alkaloid-bearing tobacco (*Nicotiana* sp.) while others were a source of dyes as was the prickly pear (*Opuntia* sp.) which, being associated with the cochinilla insect, was a source of the red pigment cochineal. Among beverage and masticatory plants were cacao (*Theobroma cacao*) which was much planted and highly prized in the lowlands of ancient Mexico and the coca bush (*Erythroxylon coca*) grown in the Andean highlands, the leaves of which are chewed to gain the effects of the alkaloid cocaine. Fruits included pineapple (*Ananas comosus*), a ground-level fruit native to South America, avocado (*Persea americana*), tomato (*Lycopersicon esculentum*), papaya (*Carica papaya*) and the breadnut (*Brosimum alicastrum*) which may have been an important staple of the Mayas. Other fruits included pepino (*Solanum muricatum*), guava (*Psidium guajava*), lúcuma (*Lucuma bifera*) and sapote (*Achras zapota*), the bark of which contains the milky fluid which is the raw material for chewing gum. It also remains a possibility that bananas and plantains (*Musa* sp.) were pre-Hispanic in parts of South America.

The majority of these plants have undergone major genetic changes in order to enhance their usefulness and although wild varieties are often identifiable, as with potatoes, the progenitors can sometimes elude attempts at discovery as witnessed by the story of maize. Some plants are little more than wild species, used perhaps as a last resort. Cañihua, for example, grown above 3,500 m (11,483 ft) in the Andes, retains its natural ability for seed dispersal and therefore must be harvested before maturity. The enormous variety of cultivated plants is undoubtedly due to the intrinsic richness of the indigenous floras, notably the equatorial rain forest, and the sharply contrasted ecological zones, which, in the cordilleran areas, are separated by very short horizontal distances. Such a juxtaposition could not fail to promote speciation and, in the human context, the exploitation of new cultigens. Variety also bespeaks a lengthy period of plant domestication.

Agricultural development and origins

The origins of plant domestication in the Americas, as elsewhere, recede into an ever more remote past with each new decade of research. Domestication and exploitation of a small number of genera has certainly been carried on from the very start of the Holocene. In the Peruvian highlands at Guitarrero Cave, common beans, lima beans and chilli peppers have been reported as early as 8500 BC while maize appears around 4000 BC. All occurences of these and other crops appear later in the Peruvian coastal sequences demonstrating later spread to this area where it appears that non-food plants such as cotton and cucurbits were among the first crops grown on any scale. In the Mexican highlands, maize appears about 5000 BC at Tehuacán while bottle gourds have been found at Oaxaca dating from about 7400 BC. These early examples, however, do not hint at one of the striking features of ethnobotany: the difference in the suite of crops grown in ancient Mexico and the Andean region. A contrast has thus been drawn between the seed croppers of Mexico, producing mainly maize, beans, squash and amaranths and the root croppers of Peru producing manioc, potatoes, sweet potatoes and pineapple. Pickersgill and Heiser (1978) point out that 'the differences in the species cultivated are the more remarkable because the dry highlands of Mexico and Peru are ecologically very similar and crops have been readily exchanged between the two areas in historic times'. The two areas are separated by forests and mountain systems and formerly perhaps, by hostile tribes. The overland distances involved favour a filtering and diffusion process rather than direct transfer since the prehistoric journey would have promised survival no more for courier than for cultigen. This situation may also suggest a strong territorial integrity among earlier societies.

There are grounds, then, for arguing a common early origin for a few selected cultigens while admitting that forest expansion may have led to increasingly isolated development in time. Crops of tropical forest origin do not appear in the records until comparatively late. Manioc first appears in coastal Peru around 1500–1000 BC. At this time contacts were made between Mexico and Ecuador and the stylistically analogous Olmec and Chavín cultures were expanding their influence from the tropical lowlands. Major cultural impulses are thus likely to have initiated trade networks between peoples previously living an isolated and inbred existence. We should also realise that the tropical lowlands were a third locus of plant domestication, indeed, agriculture may originally have spread from these lowlands to the central Andes and highland Mexico. Although the evidence for this is not particularly strong we should realize that these regions lack the intensive archaeological investigation and preserved records of the drier highlands. Furthermore it is only in the last 10,000 years that they have become so extensively forested while it would only have been possible to expand cultivation at higher elevations as temperatures rose during the Holocene.

Expansionist cultures have arisen in different ways but a very important factor seems to have been the control of water. In Peruvian prehistory, the Ceramic period, associated with the spread of Chavín influence around 1500 BC, saw a movement of population from the desert coast to the highlands. Irrigation is thought to have played a crucial part in this move while the development of pottery is obviously connected with the need to prepare, contain and cook a widening range of agricultural products. A decisive point appears to be reached when society gains mastery not simply over plant and animal domestication but over the landscape itself and begins to intensify food production around a variety of water control systems. Hence, it took a considerable time from initial domestication of maize in Mexico to real dependency upon it for subsistence. Increasingly centralized power is clearly capable of harnessing labour and resources on a larger scale and impells the twin processes of intensification and diversification. Moreover, as a better-fed population increases in size, its dependence on agriculture escalates and the society finds ways of meeting these needs or collapses. Society's capacity to meet its obligations may well be frustrated by exigencies of nature, such as climate, but as we know from modern parallels, so in the past, social forces often prove to be more influential.

We therefore recognize in the growth and decline of cultures what we may term exogenous and endogenous forces. Unfortunately, the dominant tendency in an intellectual world has been to promote, often without realizing it, a cultural materialist philosophy. The latter argues that cultural processes are merely a response to a combination of climatic, resource, population and technological constraints. Many extremely plausible arguments thus make some of man's most important discoveries and advances, such as those of plant domestication, seem almost accidental. Chance or random events, such as devastations of early Maya people by volcanic activity do fashion the course of history but one has to realize that it is environmental events and the *results* of cultural development which are preserved in archaeological sequences rather than deeper causes within society. It is now clear that cultigens were possessed by hunter-gatherer peoples at the end of the Ice Age. What does this imply for views about the primacy of sedentism in agriculture and were hunting peoples ever more than exceptionally nomadic anyway? Furthermore, in tribally-based societies the world over, secular and spiritual leaders have guided the people through time and it is reasonable to suppose that the emergence of priestly élites in later periods represents an evolution of this basic reality. The role of the shaman has been well documented and suffice it to say that a visionary or seer who was able to cure sickness, largely through an alchemical understanding of plants, would be one of the most probable originators of plant genetic manipulation. It is not as if we are without later allusions to just such a role for, while both Inca and Aztec legends speak of culture bearers (Viracocha and Quetzalcoatl), Mayan bas reliefs and frescoes depict priestly figures in association with corn plants and codices appear to record acts of planting by 'élite' figures. American anthropology and archaeology, therefore, provides us with an excellent opportunity to develop a more balanced appraisal of the whole question of agricultural origins.

Pre-Hispanic agricultural strategies

Agricultural forms and the problems they pose

A principal aim of research into early agriculture is to find out what individual areas and particular techniques produced. In many cases this is frustrated because the form of agriculture is no longer being practised or is carried on only in restricted localities making the discussion of crops and whether these were raised on a seasonal or perennial basis more speculative. We need also to address the role of former productive systems in the overall politico-economic life. Without taking account of such wider considerations, we may be short of essential explanation for the initiation of systems and for the relative care lavished in their construction and implementation. With few exceptions the examples to be discussed required a substantial labour input and, by their overall design, betoken centralized direction. At the time of Spanish contact the major powers held sway over varied ecological zones and we must therefore see the various forms of agriculture as elements within agricultural complexes, each making a distinctive contribution to the state while providing for more local subsistence and exchanges.

It is axiomatic that plots or larger-scale field systems will result from crop husbandry. The size, shape and configuration of these features relate to a combination of functional, ecological and organizational factors while their subsequent discovery depends upon agricultural subsistence having so expanded as to leave widespread remains. Ancient fields will also only be found where they have not been obliterated by later events such as are endemic to shifting cultivation and which are typified by modern commercial agriculture and urban development. It is therefore only a partial picture of pre-Hispanic agriculture which can be studied and one which relates disproportionately to marginal lands as well as to areas of more intensive research. Not all features forming integrated patterns need immediately be equated with cultivation. For example, in the highlands of Ecuador and Colombia are extensive systems of parallel alignments showing some resemblance with pre-Iron Age patterns in Britain. The latter appear to be territorial land allotments in the first instance. Such divisions will in most cases form the primary context within which cultivation occurs.

The types of agriculture shortly to be discussed represented a variety of adaptations to terrain and climate and most required periods of rehabilitation so that only a proportion of their extent would have been under crops at any one time. Furthermore, field systems represent aggregates of enterprise and for this reason also, to view them as having all been in use together would be completely erroneous. This is important to realize in the case of the many terrace systems and the emergence and decline of successive Maya cities well illustrates this process. As far as dating is concerned, fields are therefore likely to reflect their most intensive phase of use or latest remodelling rather than their initial construction.

Agriculture and domesticated animals

Discussion of agriculture sooner or later raises the question of animal domestication, but unlike the Old World there was no equivalent dependence upon domesticated animals for dairy products, draft animals were unknown, and cultivation was exclusively based on hand tools (Fig. 3.2). Bronson (1978) notes somewhat sarcastically of the Maya that 'their only farm animals were the stingless bee, the dog and possibly the ocellated turkey'. To fatten dogs on maize, as did the Maya, may suggest a degree of destitution or depravity but, although people did eat them, dogs were also used as sacrificial animals. While it is tempting to view the relative sophistication of agriculture as an artefact of the paucity of suitable animals for domestication, ·we must not overlook the fact that fish and shellfish

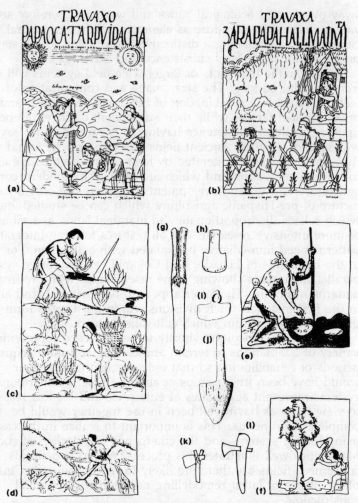

Fig. 3.2 Some early American agricultural and related implements.
(**a**) Planting potatoes with the chaquitaclla footplough (Poma de
Ayala 1615); (**b**) Cultivating irrigated maize with the liukana
(Poma de Ayala 1615); (**c**) Uictli or coa used for planting and
harvesting metl (maguey) (Codex Florentino *c*. 1570); (**d**) Uictli or
coa used for directing irrigation water (Codex Florentino *c*. 1570);
(**e**) Uictli or coa used as a planting stick (Codex Florentino
c. 1570); (**f**) Hacha (axe) used for tree cutting (Codex Florentino
c. 1570); (**g**) Socketed copper point – possibly blade of taclla-like
instrument (University Museum, Cambridge); (**h**) Copper adze
(University Museum, Cambridge); (**i**) Bronze blade of an azada
(adze); (**j**) Blade of wooden spade (after von Rosen 1924);
(**k**) Small stone hacha (axe) used for working rock (Matríacla de
Heuxotzinco); (**l**) Axe (Codex Mendocino)

were available and that in the forested lowlands there were peccary and deer whose numbers no doubt increased in appreciation of the improved diet within agricultural plots and abandoned clearings.

Coastal loci may have been more favourable for maintenance of domestic or domesticated animals in the humid and sub-humid tropics owing to the proximity of salt supplies while a similar ecological argument may be applied on the semi-arid Andean altiplanos. The central Andes, moreover, was a centre of limited animal domestication with the llama and alpaca camelids having been derived from a guanaco-like ancestor, while guinea-pigs were also domesticated, eaten and used for sacrifice. Dogs, though not native to South America, would appear to have accompanied man from the most remote times.

There is little to suggest that animals were confined in, or indeed excluded from, any of the manifestations of agriculture to be described but stone-walled, enclosed fields do exist on flat terrain in southern Campeche, Quintana Roo (Mexico) and the Petén (Guatemala) and while these might mark boundaries of ownership, they could equally well have acted to keep wild fauna from areas where vulnerable horticultural crops were being raised. It therefore seems most likely that, then as now, animals were closely associated with settlement and kept in compounds rather than in fields.

Swidden and tlacolol

Swidden or shifting cultivation occurs in the humid lowlands while *tlacolol*, an upland analogue, is, or at least, became, integrated with other more intensive methods such as terrace farming and irrigation and is associated with fixed settlement. These systems are widespread and important for their unrelenting disturbance of forest vegetation, tending to encourage rapidly-colonizing species. Though ubiquitous throughout the tropical world they were once more general elsewhere. Many have considered swidden capable of providing the surplus necessary at the time of Classic Maya ascendency even if this belief was heavily influenced by assumed modern parallels. A more modest position, according to Sanders and Price (1968), regards it as capable of supporting 'chiefdoms and non-urban states'. These systems are here regarded as the formative base on to which intensive methods were later grafted by expanding city spheres. When the latter collapsed the population became dependent on these rustic and less labour-intensive methods for, in terms of man-hours of work per kilogram of food produced, swidden is one of the most productive systems in the world. Swidden is adapted to low population densities and dispersed settlement and, in forested lowlands, provides the perfect complement to other forms of subsistence. The contemporary Machiguenga of the Peruvian montaña depend, in the words of Johnson (1983), on a 'mixture of hunting, gathering and fishing combined with cultivation of a variety of crops that provide food, medicines and materials for clothing, tools, storage containers and other manufactures'. The perceived advantage of agriculture is that it alone, in the shape mainly of starchy tubers such as manioc, can provide the storable surplus against disasters and contingencies to which isolated peoples are vulnerable. In this contemporary Indian view we therefore have a convincing motive both for the domestication and cultivation of plants.

But swidden and tlacolol have been heavily blamed for land degradation of various kinds while deforestation, not necessarily for agriculture alone, would appear responsible for thick clay layers deposited in Classic Maya times. Swidden has been condemned as wasteful and inefficient and has even excited the anger of the FAO. The trouble undoubtedly is that modern population increases, forcing people to a greater dependence on agriculture, have disrupted the traditional balance between cultivation and fallow leading to soil exhaustion. Most swiddeners approach the problem of declining soil fertility and increasing weed problems by planting first the most demanding crop, usually maize, and it is for this reason that swidden gained its Mesoamerican name of milpa or corn plot. Swiddens are variable in the detail of their management; in drier areas burning is no problem but two crops a year may not be possible while in wetter areas initial preparation of sites may be more fraught yet continuous cropping is possible. Tubers are normally grown in raised beds while various forms of under-planting and inter-planting are usually carried out, such as growing mixtures of corn, beans and squash which help to conserve soil fertility, control weed growth and appear to maximize yields. There is no reason to suppose that this situation was any different in pre-Hispanic times.

Tlacolol is widespread on slopes in the humid parts of Mesoamerica and Andean South America. It involves a shorter fallow cycle than swidden and traditionally makes use of a hoe (*coa* or *uictli*) to prepare land for sowing and for weeding. The land is divided into sectors some of which are planted for two to three years then left fallow for periods of up to seven years. The length of the cycle depends on inherent soil fertility – many are based on fertile volcanic rocks – as well as humidity and the depradations of plant parasites and pathogens. The ratio of land lying fallow to that cultivated may therefore be as low as 1 : 1 but more commonly is around 1 : 3 comparing with true swidden where an optimum ratio might be around 1 : 12. Although tlacolol may once have been a completely free-standing system, it became in time part of an infield-outfield system, with permanent, intensively cultivated lands around the settlement, nearer to irrigation water and sources of domestic refuse. It is in this broader ecological and community context that tlacolol has sustained population growth probably beyond the population densities of swidden.

Hoe cultivation of suitable freely-draining slope soils appears to have expanded in parts of the Andes from about 4000 BC while since

Plate 2   Turning the soil on fallowed land, near Copacabana, Bolivia (1971).
Two men working with footploughs (*tacllas*) dig in unison while a
woman turns the sod that they have broken with her mattock (hoe).
(Photo: D. A. Preston)

about 1500 BC stone hoes, suitable for cultivating root crops are
found mainly at higher elevations. From this time maize, beans and
squash were more widely produced at lower elevations. In parts of
Central America at Spanish contact the drier, forested slopes were
the main sites of agricultural development while the lower, wetter
parts of valleys were open and periodically burned, possibly to
encourage grazing animals. Similar patterns are discernible in the
Peruvian montaña. The parallel land divisions widespread in parts of
the northern Andes, while bespeaking an equable division into
allotments may also fit the sectoral model of tlacolol and the
tendency for early agriculture to avoid bottom lands. It is possible
that these features represent a prototype of the Andean sectoral
fallow system known as *laymi*. Fire can more easily be controlled
burning up-slope, which is the most frequent practice, so the logic
of narrow, parallel, up-and-down-slope strips becomes clearer.

Swidden and tlacolol are not equal to generating a large surplus
but more can clearly be squeezed from them for short periods. Their
role at times of centralized power in ancient America can only have
been extremely peripheral though the Peruvian montaña and the
Gulf Coast lowlands clearly did provide exotic produce for Cuzco
and Tenochtitlán. Nevertheless it is a matter of some speculation as
to what the effect of agricultural intensification was on these native

systems. The latter may have led to such a withdrawal of labour as to induce abandonment of marginal lands while, locally, increased demands on these fragile systems might have had disastrous ecological and political consequences.

Terrace farming

Agricultural terraces represent a truly astonishing investment of effort by aboriginal peoples in the New World. They signified the growth of agriculturally dependent communities and are therefore to be seen as a form of agricultural intensification. We assume that the benefits were not simply increased overall production but larger and less fluctuating yields with the possibility this allowed for wider exchanges. They were also a context in which to build and maintain soil fertility and reduce the length of fallow periods although many terraces in the driest areas would only have been used for annual crops in the wetter years. In the more humid areas terracing may have developed in response to long-term problems of managing tlacolol. Characteristic of arid and seasonally dry areas, terraces were adapted to irrigation, the history of which closely parallels terrace construction from the first millennium BC. Significantly, there was also an expanded development of pottery and metalworking from this time. Nevertheless, many areas of terracing are too high and remote from water sources or are constructed in a manner which otherwise militates against effective irrigation.

By far the most widespread are the many kinds of contour bench terrace, usually stone-faced and sometimes with elaborate foundations (Fig. 3.3). On gentle slopes terrace surfaces often maintain a slight downhill gradient while on steeper slopes the planting surface is often nearer the horizontal. On coca terraces, which are very narrow, planting takes place in a trench which runs along the backwall. The terrace form not only stabilizes soil but improves water absorption and retention, thereby maximizing meagre rains. Terraces also contribute to a more uniform drainage condition along contours than normally obtains in nature. Terracing at higher elevations also compensates for problems of frost drainage experienced on valley bottom sites. In deep gorges, terraced slopes can also provide favourable insolation as at Inty Pata (= sunny slope) in the Urubamba Valley, Peru. Terraces (*trincheras*) were also commonly developed in Mexico across the channels of ephemeral streams, the stone structures serving as check dams. This facilitated the accumulation of deep, water-retaining soil. Close relationships therefore can exist between terracing and irrigation, just as they do between irrigation and drainage, which warns against attempts to compartmentalize rigidly these various agricultural strategies.

Terraces were one component, rarely the only component, in the local agricultural economy and one reason for their widespread distribution was that their construction and farming required co-operation at no higher a level than that of the village community, in

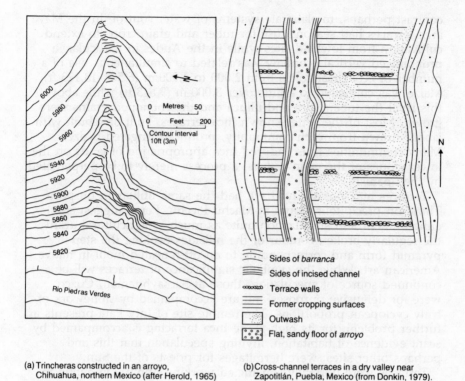

(a) Trincheras constructed in an arroyo, Chihuahua, northern Mexico (after Herold, 1965)

(b) Cross-channel terraces in a dry valley near Zapotitlán, Puebla, Mexico (from Donkin, 1979).

Sides of *barranca*
Sides of incised channel
Terrace walls
Former cropping surfaces
Outwash
Flat, sandy floor of *arroyo*

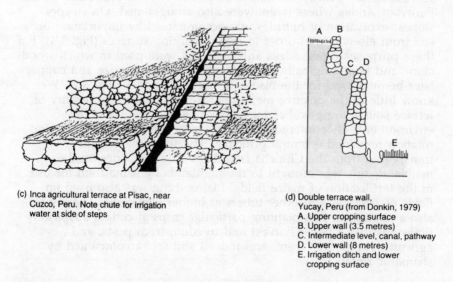

(c) Inca agricultural terrace at Pisac, near Cuzco, Peru. Note chute for irrigation water at side of steps

(d) Double terrace wall, Yucay, Peru (from Donkin, 1979)
A. Upper cropping surface
B. Upper wall (3.5 metres)
C. Intermediate level, canal, pathway
D. Lower wall (8 metres)
E. Irrigation ditch and lower cropping surface

Fig. 3.3 Agricultural terraces in Mexico and the Andes

contrast perhaps, to elaborate systems of water control. In the Maya area terraces had seasonal use for tuber and grain crops to extend cultivation from lowland sites while in the Andes in particular, a pronounced vertical economy was related to simultaneous use of a number of ecological zones from 2,000 to 4,000 m (6,562–13,123 ft). Maize was cultivated to a little over 3,000 m (9,843 ft) while above this level the tubers, oca and ulluca and the chenopod grains were grown. Most of the better-known Inca terraces, built or rebuilt with labour drawn from each community, produced irrigated maize. Many communities thus endured Inca appropriation of lands on the lower slopes of the main valleys, a process merely to be perpetuated by the Spanish.

Terraces may not always have had the same utilitarian purpose. Those on precipitous slopes at Machu Picchu, Peru, suggest site enhancement and stabilization quite as much as agricultural utility, the regularly broken contour of the hillsides evoking the step pyramid form and corresponding to an oft-repeated motif in native American art. While the symbolic significance of terraces will be a continued source of speculation, those at Sacsayhuamán, Cuzco, were for defensive purposes and are accompanied by stonework of truly cyclopean proportions. The remote site of Inty Pata presents a further problem, for its high-grade Inca terracing is accompanied by scant evidence of habitation, inviting speculation that this and perhaps other sites, were hermitages for priests of the Sun.

Terraces were normally constucted working uphill. On gentle to moderate slopes the dry stonewall merely prevented uncontrolled downward movement of soil and water. For example, outwash fans were modelled by excavation and forward filling to produce a succession of virtually flat surfaces such as at Pisac and Yucay in the Peruvian Andes where rivers were also straightened. On steeper slopes, excavation of hillsides was accompanied by importation of soil from elsewhere in order to create planting surfaces (Fig. 3.4). For these purposes picks, adzes and mattocks were used in which wood, stone and bone originally predominated, though bronze and copper later became used for the blades of hoes and adzes (Fig. 3.2). We know little of the positive measures taken to ensure the fertility of terrace soils – these will certainly have been as varied as the environments encountered – but we can be sure that fallowing and rotation remained a central principle. The Incas had guano transported from the Chincha Islands by raft and, once on the mainland, this was brought to the highlands by llama train for use in the fertilization of maize fields. Llama dung was also used on fields of potatoes and other tubers at higher elevations. There was also a realization that planting particular crops at certain auspicious times ensured a better harvest and freedom from pests, and agricultural practices were, and indeed still are, accompanied by shamanic ritual.

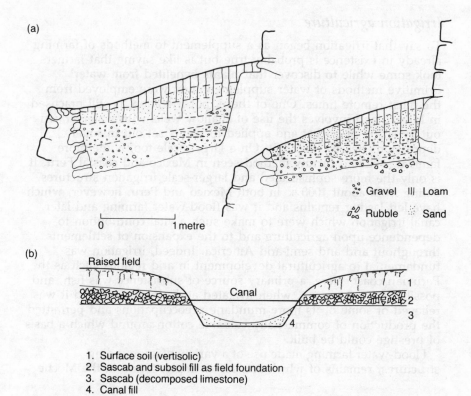

(a)

Gravel | III Loam

Rubble | Sand

0   1 metre

(b)

Raised field

Canal

1. Surface soil (vertisolic)
2. Sascab and subsoil fill as field foundation
3. Sascab (decomposed limestone)
4. Canal fill

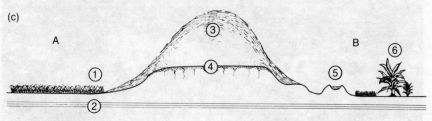

(c)

1. Field surface of Distychlus grass and salt crust
2. Ferruginous layer marking position of ancient water table
3. Mound of biological and mineral debris from excavation and clearance of field surfaces
4. Old land surface, usually indistinct
5. Small canal for irrigation
6. Field crops of corn, sweet potatoes, squash, beans and locally small fruit or figs

Fig. 3.4 The disposition of soils in pre-Hispanic terraces, raised fields and sunken fields. (a) Soil cross-sections through terraces at Patallacta in the Cusichaca Valley, tributary of the Rio Urubamba, Peru (after Keeley 1985); (b) Section of Maya canal and raised fields, Pulltrouser Swamp, Belize (after Turner and Harrison 1981); (c) Composite drawing of abandoned (A) and currently operational (B) sunken fields at Chilca, Peru (Author)

Irrigation agriculture

To say that irrigation began as a supplement to methods of farming already in existence is probably true but is like saying that farmers took some while to discover that plants benefited from water! Primitive methods of water supplementation were employed from the most remote times. One of these, pot irrigation, is still practised in Mexico and involves the use of shallow wells with water drawn out by scoop or bucket and applied to plants set out in small depressions nearby (Fig. 3.5). On a larger scale too, is the more familiar well and furrow method seen in Mexico and coastal Peru. It is only the more sophisticated and larger-scale irrigation structures, dating from about 1000 BC in both Mexico and Peru, however, which have left lasting remains and it was flood-water farming and later, canal irrigation which were to make such a vital contribution to dependence upon agriculture and to the expansion of settlements throughout arid and semi-arid America. Indeed, irrigation was fundamental to agricultural development in arid regions such as the Peruvian coast. Here, a primary source of subsistence was fish, and possibly for this reason, when irrigated agriculture expanded it was relieved of some of its more mundane preoccupations and permitted the production of commercially valuable cotton around which a basis of prestige could be built.

Flood-water farming made use of a variety of impounding structures, remains of which are, for example, found in the Moche

Plate 3   Ancient raised fields in Central Veracruz, Mexico. (Photo: A. H. Siemens)

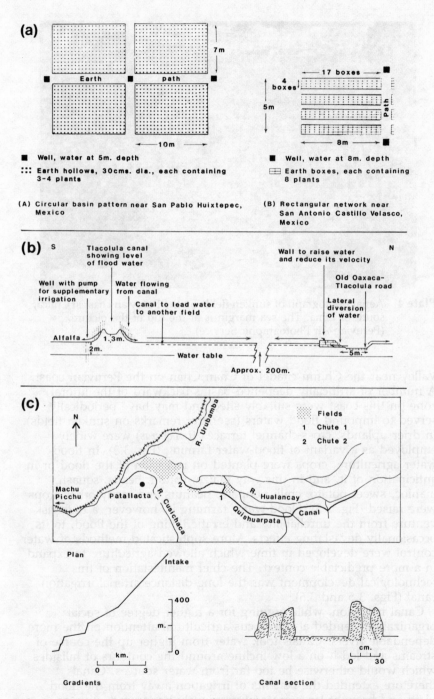

Fig. 3.5 (a) Pot irrigated fields (after Kirkby 1973); (b) Cross-section of floodwater-farmed area near Tlacolula, Mexico (after Kirkby 1973); (c) Quishuarpata Canal, Peruvian Andes (after Farrington 1980)

Plate 4 Aerial photograph of sunken fields on the Peruvian coast at Chilca, south of Lima. The sea margin is at the top of the picture. (Peruvian Air Photographic Service)

Valley near the Chimú citadel of Chan Chan on the Peruvian coast. A number of artificially deepened basins backward of the littoral zone on this coast were suitably sited and may have periodically served to impound flood waters (see later remarks on sunken fields). In drier uplands, cross-channel terraces (*trincheras*) were widely employed as a variant of flood-water farming (Fig. 3.3). In flood-water agriculture, crops were planted on recession of the flood or in anticipation of it, and in this way maize, cotton, beans, squash, manioc, sweet potato, chilli peppers, peanuts and many other crops were raised (Fig. 3.5). Flood-water farming is, however, a high-risk venture from the unreliability and erratic timing of the flood, to its occasionally devastating effects. More sophisticated methods of water control were developed in time which allowed agriculture to expand in a more predictable context. The chief manifestation of this technological development was the long-distance arterial irrigation canal (Figs. 3.5 and 3.6).

Canal irrigation, while arguing for a higher degree of social organization, tended also to focus agricultural attention on the more dependable rivers. Canals took water from higher up the course of streams and led it on a low incline around the contours of hillsides which would otherwise be too far from water sources. Canals therefore extended the benefits of irrigation away from the flood plain and brought large areas of previously underproductive or uncultivated land into production. It is easy to see why canal irrigation so often went hand in hand with terrace construction.

Many of the more sophisticated terraces were provided with sub-wall drainage and spouts allowing water to pass harmlessly to lower levels; both are associated with Maya and Inca works (Fig. 3.3). Arterial canals sometimes depended on major engineering works and at Copán in Honduras are the remains of a Maya dam while reservoirs and associated canals occur at Edzna, Campeche, in Mexico.

Canals were also an essential feature of the drainage and integrated management of tropical wetlands in pre-Hispanic times (see discussion of drained fields) so we should view their advent for both drainage and irrigation as having an importance greater than simply the emancipation of agriculture. In reality the controlled abstraction of river water and the principle of the canal permitted the flood-plain environment to be tamed, with considerable implications for the development of settlements and communications.

Drained fields

Drained, and most usually, raised-bed fields are to be found in three general localities; the moist tropical lowlands, the highland lake basins of Lake Titicaca and the Valley of Mexico and sloping terrain in the central and northern Andean highlands. Many variations are encountered from flat planting surfaces divided by drainage ditches or channels (ditched or channelized fields) to those elevated in different ways by construction (raised or ridged fields). Thus, ridges and platforms of many sizes and shapes, often forming intricate patterns, were created to produce what were essentially drained fields (Fig. 3.6). The significance of this kind of field is that in areas of markedly seasonal rainfall one of agriculture's orientations is towards maximizing use of the landscape in the one or more wet seasons while making a deliberate attempt to exploit wetlands for dry season crops, in which mode, drainage ditches can become a source of irrigation water.

The creation of ridges and raised platforms has a number of potential agronomic and microclimatic advantages. In semi-arid regions it will facilitate concentration of salt above the level of plant rooting, one of the roles which ridges probably performed in pre-Hispanic times in the Lake Titicaca basin. Pronounced ridging can generate shelter indirectly by interference with airflow and helps to control soil drift. It may, in particular instances, also assist in frost protection, all of which helps explain the occurrence of pre-Hispanic raised garden beds not only in the American tropics and high Andean altiplano but in the Great Lakes region. The most obvious value of raising the planting surface, however lies in drainage and the optimization of soil biological processes in terms of temperature and aeration. Crop production would benefit from such a strategy in lowland and upland areas alike.

The Maya developed drained field systems, many of which first

53

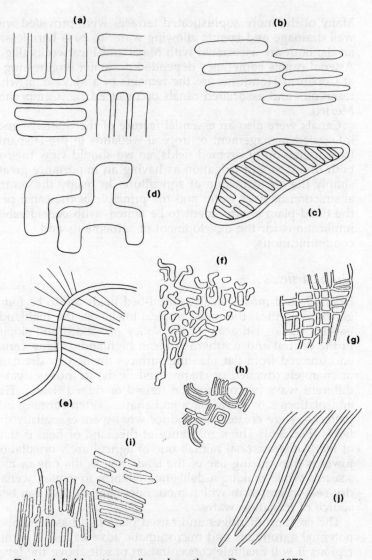

Fig. 3.6(A) Drained field patterns (based partly on Denevan 1970;
Denevan and Matthewson 1983 and Siemens 1982). (a) Regular
checkerboard pattern; Chinampas, Valley of Mexico and parts
of Titicaca basin; (b) Ladder or echelon formation, e.g. Titicaca
basin, Guayas basin, Ecuador; (c) Embanked systems, e.g.
Llanos de Mojos, Bolivia; San Jorge, Ecuador; (d) Interlocking,
e.g. Vera Cruz, Mexico; also Belize; (e) Ditched fields in river
levée zone, e.g. Campeche and Vera Cruz, Mexico;
(f) Irregular platform fields, Guayas basin, Ecuador;
(g) Ditched fields with canal network, Guayas basin, Ecuador;
(h) Platform fields, Samborodón, Guayas basin, Ecuador;
(i) Linear ridged fields, Guayas basin, Ecuador; (j) Highland
ridged fields, northern Andes. Form very variable.

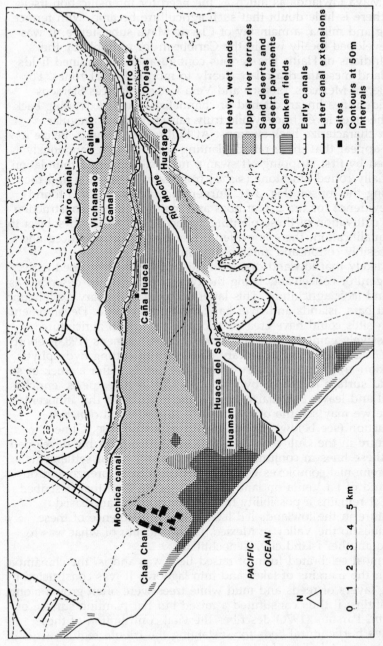

Fig. 3.6(B) The physiographic context of canal-irrigated lands and sunken fields at Chan Chan, Peru (after Farrington 1974)

appeared in the Classic period (AD 400–800). The increase in their extent at this time may have reflected the deteriorating soil drainage of the Maya lowlands as much as the need for intensification itself while there is little doubt that such agriculture became, with terrace farming and milpa, a mainstay of Classic Maya subsistence. It was also developed locally within the Caribbean for example by the Taino Indians of Haiti. The various configurations of drained fields in lowland mesoamerica relate clearly to morphological zones. For example, in Mexico's Tabasco and Vera Cruz provinces Siemens (1983) identifies linear, ditched or channelized fields extending back from the river levée tracts while truly raised fields of irregular pattern characterize back swamps. At Pulltrouser Swamp, Belize it would appear that these raised planting surfaces were laboriously prepared by first stripping off swamp materials, laying a foundation of partially rotted limestone (*sascab*) from canal excavation and then returning the topsoil (Fig. 3.4). Turner and Harrison (1981) found some evidence for the use of hoes and mattocks in the construction and/or cultivation of these planting surfaces and, although subject to some qualification, there is evidence that maize was produced and also possibly cotton and amaranth grain.

It is likely that lowland drained field cultivation developed early, its progenitor conceivably being the type which may be described as *playa* and *levée* agriculture. This latter involves temporary exploitation of sand bars, islands and river banks as floods recede. Development of such sites could have created a locus from which agriculture progressively gained control of back swamps. Siemens (1983) reports the practice among contemporary Tabascans of cutting hydrophytic vegetation and planting maize among the debris. After a week or so the field surface is cleared by burning which reduces pests, enriches the soil and leaves the maize largely undamaged. In this innocent practice we may see the origin of more sophisticated *chinampa* construction (see below). In the second millennium BC, Olmec agriculture in the Gulf coast lowlands of Mexico probably developed along these lines in conjunction with milpa, for major settlements and ceremonial complexes were developed at San Lorenzo and on the island of La Venta on land no more secure than that described above. It remains a possibility that adoption of aquatic-based agriculture in the lowlands led later to the transference of these techniques to the Valley of Mexico, to the vicinity of what was to become another island city, Tenochtitlán.

The most celebrated form of raised field was and is, the chinampa, built on the margins of lakes and into lagoons. It was constructed by placing layers of reeds and mud while trees were often grown along the margins; it thus constituted a raised but flat planting surface or platform. Parsons (1976) describes the vital contribution of these intensive horticultural beds for sustaining the tribute requirements of the Aztec capital, Tenochtitlán, now the centre of modern Mexico City. The latter lay in brackish lake Texcoco while to the north and south in fresh water at higher levels, chinampas were developed, particularly in the Xochimilco-Chalco lake bed where it is estimated they once covered up to some 9,500 ha (23,475 acres). A considerable expansion occurred in the period 1426–1467, mostly to provide maize. Here, chinampas were of vital importance owing to the aridity of the climate and the vulnerability of rain-fed farming.

It is understandable to regard swamps and aquatic habitats as marginal for agriculture, yet we do well to note that the biological productivity of marshes and swamps can be up to two and a half times higher than forest ecosystems and perhaps five times higher than conventional cultivated fields. The agricultural potential of reclaimed wetlands is very high indeed. Not only are the beds seasonally flushed with nutrients but organic, nitrogenous material can be dredged from adjacent ditches and canals to build and enhance soil fertility. It can scarcely be argued that, with primitive tools, soft muds and sand presented any more awesome a task than did breaking into hard dry turf, while the rivers and canals around the fields were a source of fish and a variety of other protein. Semi-aquatic drained fields have arguably the greatest capacity of any single agricultural system to sustain production year after year in marked contrast to milpa and other more extensive, rain-fed systems. In this respect, nutritionally they can resemble contemporary rice paddies, although cultivated rice was unknown in pre-Hispanic America. Tropical lowland drained fields grew many of the crops of the rain forest such as manioc and, as drainage systems became more sophisticated, as in the Maya lowlands, year-round cropping would have become possible, with the spatial and temporal water regime determining the pattern of crops. As Siemens (1983) comments, raised fields may have provided 'a seasonal complement to the wet season milpa and other subsistence activities' while in areas where fields were, or could be made to be, above flood levels 'the system might also have been the crucial element in the support of nearby population concentrations'.

In the central and northern Andes, ridged or ditched fields cover very large areas, with notable concentrations occurring in the highlands of Ecuador and Colombia, in the semi-arid Titicaca basin of Peru and in the seasonally inundated Bolivian *llanos*. These features are generally at least 2 m (2.2 yards) in width and distinct from the much smaller individual soil ridges associated with the planting and ridging of rows of root crops. The form of ridged fields shows variation regionally and their widths and individual topography also reflect to some extent the angle of slope of hillsides. The object of these fields appears to have been the expansion and intensification of wet-season cultivation mainly of the eminently storable roots and tubers, the latter having been grown up to heights of 4,500 m (16,403 ft). At the higher elevations potatoes, quinoa and cañihua were the main cultigens whilst greater diversity would have been encountered at lower altitudes with maize and the coca bush being widely grown. In this context as elsewhere, the diversity of plants would have been greatest around settlements with herbs for medicinal or condiment purposes being grown as garden crops. The instruments used for field construction and cultivation are likely to have been forms of the *coa* and *chaquitaccla* (Fig. 3.2), the latter so-called foot plough having been clearly designed for breaking through dry soil which has developed a turf cover. Varieties of mattock (*raukana*) would have been of particular value and an historical account from the Orinoco llanos of Venezuela refers to construction of ridged fields with what were called *macanas*. In any event the widespread regularity of field strips suggests that teams of labourers worked to pre-arranged plans determined by principles of land division as well as by topographic constraints.

The obvious contrast with terrace agriculture is that in areas subject to wet seasons and tropical storms, or simply a high water-table, a moisture-shedding rather than a moisture-conserving role is the priority and this, irrespective of the duration of the wetness. It was stated earlier that terrace farming was associated chiefly with moisture-deficient areas and that it probably superceded tlacolol systems. Much the same temporal relationship may have applied in the case of ridged fields. One may suppose that agricultural intensification in the wetter areas likewise involved the exploitation of increasingly steep slopes on which primitive sectoral agriculture would have tended to invite soil instability, from soil wash to larger scale earth slips. Such risks do of course depend on a variety of circumstances, important among which are the depth and texture of the soil and the permeability of the subsoil and subjacent rock. Hence, a view recently put forward is that up-slope ridged fields were designed to circumvent slope failure. The combination of ridge and ditch facilitates surface run-off and generally improves soil drainage. In particular circumstances, as where ash deposits overlie impermeable subsoils, ridging increases the capacity of the cultivated areas to absorb moisture while drainage lines permit excess water to drain from above the impermeable base, thus reducing or even avoiding the risk of soil slips through lubrication.

The fact remains, however, that slope soils are frequently lacking in depth, a problem which is likely to have been perceived as limiting in the case of roots and tuberous crops. Measures taken to thicken the soil artificially would improve productivity notably through increasing the amount of water and nutrient accessible to the growing crops. Furthermore, any tendency, however brief, for saturation to occur would be confined to levels below the main rooting zone. There can be little doubt therefore, that ditching and ridging conferred both stability and agronomic benefits to the cultivation of sloping terrain in wetter areas of tropical America. Ditching was probably always the primary objective. Ridging in the highlands sometimes depended upon topographic circumstances, as in the lowlands, but was probably always mainly a function of field width and soil type. The latter are likely to have been interrelated. For example, the less able was the ditch to maintain cut sides, the broader it may have become and the more soil may have been placed onto the field surfaces.

The up-and-down-slope orientation of so many ridge field systems might have seemed certain to lead to soil loss and gullying through excessive, rapid run-off along the inclined drainage lines. It certainly did so in some localities yet the system is so widespread that one is forced to conclude that it was basically sound. For example, it is now clearer to see why, had these fields normally been arranged in bunded fashion around the contours, serious buildup of subsoil water could occur, thus destabilizing large areas. It needs also to be appreciated that, following initial excavation of the ditches there will have been a natural tendency for these to become overgrown with a durable turf dominated by grasses. Initially this process will have taken place during the drier intervals. The latter process recalls the use of sod channels and terraces in modern soil conservation practice. The greatest run-off would always have occurred when the first heavy rains encountered hard, dry soils; thereafter, the soils would improve in their ability to transmit and to hold moisture. Additionally, cultivation almost certainly did not take place at all points along each ridge at any one time and would have been confined to the convex or flattened top of each ridge if only to allow better access for weeding and harvesting. It seems equally probable that rows of plants were cultivated at right angles to the orientation of the field, i.e. with the contour. Such soil management would again have helped reduce down-slope soil losses.

Although these explanations enable us to see why inclined field systems were viable and have become such an enduring expression of pre-Hispanic agriculture there are dangers in assuming that they are applicable to every ridged field group or that pre-Hispanic peoples made their decisions exclusively in response to physical processes. These fields represented the distal portions of divisions of land which were related to settlement, mostly of valley rather than hillside location. We are thereby compelled to see the logic of an up-and-down arrangement as it relates both to access and to the management of individual units of the pattern.

As one examines further the context and relationships of drained fields it is possible to see how sloping field agriculture was complementary and gave way to the swidden type of agriculture in the humid lowlands; how it was replaced locally by irrigated terraces and by terrace farming and tlacololes generally in the drier regions. The reconstruction of entire pre-Hispanic landscapes becomes, therefore, a meaningful and realistic objective for even where field systems are not in evidence the soils near to concentrations of settlement in both temperate and tropical areas are found to be characteristically 'anthropic'. In the Amazon headwaters region of Colombia, for example, sites on low terraces which are close to and have good views of navigable waterways are often characterized by their organic, *terra preta* soil horizons. Here, the benefits of free drainage are apparent and such sites were clearly the sophisticated tribesman's answer to playa and levée agriculture. Riverine sites and those elevated above rivers, especially river bends, were not only the loci of tropical lowland settlement, they were the typical settlement sites for the Adena and Hopewell and later Mississippian woodland peoples of eastern North America whose original resource base could be very favourably compared with that of the tropical lowlands.

Sunken fields

The most geographically restricted form of indigenous agriculture is that of the so-called sunken fields of the desert coast of Peru. These consist of elongated depressions excavated in sand and their close association with a fresh water-table suggested that they were dug to exploit the capillary fringe (Figs. 3.1, 3.4 and 3.6). However, I have argued elsewhere that the system would not have been viable without use of surface-applied water derived from wells. The few sunken gardens which remain in use for crop growing at present, notably at Chilca to the south of Lima, rely on irrigation water pumped from wells and this must in any case have formerly been the local source of domestic water. In addition, there are grounds for believing that some sets of integrated basins were suitably located for receiving flood water. This has led to some controversy which may be resolved in terms of the evolutionary history of these sites and of the coastal plain as a whole. The basins are quite distinct from walk-in wells or sites associated with simple 'pot' irrigation and are commonly surrounded by steep-sided mounds of mineral and organic refuse from field surfaces. Nevertheless, fig trees are today grown in individual pits which reach towards a moist zone. Sunken fields formerly produced crops for food and industrial purposes, many of which are inherently salt-tolerant or suited to a high water-table; these include beans, maize, sweet potatoes, peanuts and a range of other vegetable and fruit crops. There are also abundant remains of cotton, gourds, squash and reeds, the latter being grown to this day in brackish, water-filled depressions. These were, and to a limited extent still are, the raw material for boat-making, and for house construction in association with adobe.

Various attempts have been made to place sunken fields into a spatial and community setting. At Chan Chan, at the mouth of the Rio Moche, sunken gardens were clearly tributary to a major dependence on irrigation farming associated with the Imperial Chimú domain. On the other hand, at Chilca there is no equivalent urban development and here, the large extent of sunken fields relate to the ephemeral nature of the Chilca stream. The field systems at Chilca have very largely been constructed where the dry Chilca river mouth reaches, but fails to cross, the littoral zone. Although floodwater is said to be available on occasion, it seems reasonable to surmise that where surface water supplies were unreliable, ground water-based agriculture would have tended to proliferate. Why so large an extent of sunken fields was developed at Chilca is intriguing and it would seem highly probable in view of traditonal contacts between highlands and coast that this development and others like it, took place at the instigation of highland chiefdoms anxious to broaden and strengthen their resource base. The coastal sites provided abundant fish and other materials including guano. It is the writer's contention that such forms of subsistence were not initiated by external influence but their organized expansion – so evident at Chilca and elsewhere – resulted from Tiahuanaco and Huari highland expansion in the early centuries of the Christian era, such a timing being borne out by recent radiocarbon dates.

Post-conquest changes and the case for traditional agriculture

The spread of Spanish rule throughout the greater part of Central and South America caused considerable upheaval. First and foremost, the indigenous populations declined dramatically, not least through succumbing to the Europeans' diseases such as the common cold and those of childhood. Under circumstances of abject native debilitation the Spanish were able more readily to exploit their new territories and establish a new religion and characteristic urbanized life. As with the spread of the Roman empire, this process was facilitated by the existence of a developed agriculture and a system of established portage routes. Large-scale abandonment of native agricultural lands was to result, this being reinforced by policies of *reducción* or *congregación* in Peru about 1570 and in Mexico after about 1590. Abandonment of terraces led in some cases to their degradation and eventually to gully erosion. The introduction of plough technology saw the rapid extension of agriculture into moist but fertile valley sites, many of which had previously been beyond the scope of hand tools. In these locations *haciendas* typically were established which had the effect of forcing native agriculturalists to make do with the steeper slopes and more remote areas at higher elevation. The Europeans introduced a range of exotic crop plants to different parts of tropical America including date palms, citrus, figs, grapes and wheat. Exchanges became more frequent between North and South America and the dessert banana and cooking plantain became widely grown. Later the eucalyptus and other exotic trees made their appearance. The horse, which had so terrorized the defending Aztec and Inca warriors was soon joined by its small cousin the donkey, together with cattle, goats and pigs. While the donkey largely replaced the Andean llama as a general purpose delivery van, the introduction of new grazing animals was to have major implications for the provision of pasture. This also contributed to a restriction of indigenous subsistence to *tlacololes* on the steeper slopes and it is probably from the sixteenth century that a major decline in woodland resources set in. Indeed, recent increases in the extent of tropical grassland are a further illustration of the role of animals in landscape change and it is to be wondered to what extent the grazing of native grasslands and scrubland in the colonial period by exotic cattle and goats, brought about environmental deterioration, particularly in drier areas.

It would be wrong to assume that prior to European contact there had been centuries of virtual homeostasis in the forms and products of agriculture for, in terms of potato culture at least, the situation appears to have been one of continuous selection and change. Nevertheless, at the present time, economic imperatives manifesting in scale of enterprise and choice of crops, threaten to undermine indigenous farming as never before. In the first place there has been a marked increase in the scale of commercial farming or agribusiness from bananas to hennequen and from sugar cane to cotton. This has led to the extermination of native agriculture either by taking its land, its people or both. The transition some twenty years ago in a Panamanian village from a subsistence rice and maize system to cash cropping of sugar cane has been described. While the change is basically an ecological one, we must also be aware of the change in attitude among *campesinos* which accompanies the changing type of production. It is clear that changes of this sort are even more irreversible than are those of ecology. Such developments in the tropical lowlands have at the very least led to pressure on the remaining peasant lands, the response to which has been the adoption of permanent settlements with resulting land degradation. In recent years it is the large who have gained at the expense of the small at all the various scale levels in farming, a situation so well encapsulated by the maxim 'Get big or get out'.

Advances in plant breeding associated with the so-called 'green revolution' only seem to have endorsed this situation for, as Greenland (1975) has observed, 'the technology required for high yields – fertilizers, pesticides, machinery for tillage and harvesting, proper irrigation control and access to markets – is not available to the small farmer nor (in developing countries) is it adapted to his level of education and normal scale of operations. Furthermore, the small farmer usually does not have the capital or access to credit needed for this type of operation which in any case does not give him any greater guarantee of a minimum, secure level of production'. The introduction of higher yielding varieties into areas of indigenous subsistence farming, though it may improve the economic prospects for well placed campesinos, also carries its own set of problems as Andean potato culture illustrates. Until recently, approximately two hundred varieties of indigenous potatoes of various land races were in use throughout the Andean potato heartland. At the present time, however, improved hybrid varieties are grown increasingly as cash crops, are two to three times more productive than native varieties, are resistant to potato blight and respond well to fertilizer. This trend inevitably affects patterns of exchange established over many hundreds of years and it is claimed that 'genetic erosion' or loss of genetic diversity is taking place rapidly and is most serious on the margins of the potato heartland. The risk is not that native varieties will disappear altogether, for people living in the Andes generally prefer eating the tastier indigenous types to the introduced ones and they also store better; it is simply that many different races of *Solanum* are suited to different land types and should failure attend either the introduced potato or its market, the native *campesinos* would no longer have suitable alternatives on which to fall back. It is not only potatoes which are affected in this way in the Andes for wheat and barley are spreading in popularity for similar reasons, threatening especially the indigenous chenopods, quinoa and cañihua. Likewise, European broad beans (*Vicia faba*) are seriously displacing the indigenous *frijol común* (*Phaseolus vulgaris*) and tarwi (*Lupinus mutabilis*) in the central Andes. It has furthermore been traditional to grow potatoes for up to two years, followed by a short rotation of crops such as oca (*Oxalis tuberosa*) with chenopod or amaranth grains but not returning to potatoes for at least six years in order to avoid eelworm infestation. Interference with individual crops or the rotations as a whole obviously will have important consequences.

So must we be faced with the relentless tendency towards monoculture and should we be content to merely record the passing of native agriculture as today we also witness the inexorable elimination of the rain forest? Such conservationist questions are perhaps symptomatic of those distanced from the reality and enjoying superior material living standards. Indeed it would be morally reprehensible to suggest on purely conservationist grounds that farmers ought to continue with practices which currently register as bare subsistence. It is however, equally the moral responsibility of those able to foresee agro-ecological crises to help others plan their development in a way which does not compromise the future. In the face of contemporary trends there may appear to be no substantial case for preserving traditional farming. Indeed, the plural society (and largely dual economy) has ensured native agriculture would never be more than an economic backwater. It is well to realize, however, that the major achievements of indigenous farming, in terms of monumental remains, are as much a testimony to the coercive nature of pre-Hispanic societies as to their technological and agronomic prowess. Indigenous peoples may now be subject to cultural erosion but, despite hardships, have arguably gained a freedom of action which did not exist in former times and it is this which now constitutes an important factor in the future viability of traditional agriculture.

If we argue for the preservation of indigenous agriculture we will probably have in mind the fact that this alone appears able to utilize a range of extraordinarily hostile and isolated terrain and the fact that modern technology has little place among lowland rain forests or over vast tracts dominated by steep slopes. In this respect countless numbers of holdings in the American tropics are below 2 ha (5 acres) and as energy costs continue to rise, mechanization, except in terms of intermediate technology, appears increasingly doubtful. Under these circumstances, with access to markets being a crucial factor, self-sufficiency and local exchanges can be presented as a logical solution. It should also be recognized that small horticultural plots are proportionally far more productive than large-scale mechanized systems.

As indigenous farming and farmlands are increasingly abandoned we should try to identify in what respect the loss is most serious. Regrets about the possible under-use of land are valid, sentimentality understandable but the passing of rich ethno-botanical inventories and agricultural traditions is a serious loss to humanity as a whole. Indian practices are adjusted to maintaining the fertility of the land and to maximizing yields albeit of indigenous crops. The practices of planting mixtures and of adopting particular sowing calendars, e.g. according to the moon's phases have much in common with Bio-Dynamic agriculture, a truly holistic form of ecologically orientated farming. These are ultimately the most tragic losses and in this author's view future study should be directed to the latter area so that, at the very least, indigenous practices may beneficially infuse themselves into contemporary farming systems. It is encouraging to note that a training centre for sustainable agriculture has recently been established in the Dominican Republic.

Indigenous systems and the future

The case for indigenous farming is then based on its careful adaptation to environment and its maintenance of biological diversity and cultural stability. Its methods are relevant to a world of declining resources and to increasing costs of raw materials, particularly energy. Although associated with low per capita output, indigenous systems are nevertheless labour-consuming rather than labour-shedding. For this reason, especially when world markets are in recession it should be the aim of national governments to stem the tide of *campesinos* (small scale farmers) moving away from rural areas to what they so tragically see as a better life elsewhere (see Ch. 9). Traditional rural agriculture produces food for the family and community while modern mechanized agriculture is geared to the urban and overseas market place. The fact is that rural populations over wide areas, together with workers actually employed in commercial agriculture, lack the income to buy food produced by this expanding agricultural sector. The problem is highlighted by the production of grain and the irrigation of pastures for the purpose of feeding high-value livestock rather than people. In the face of these problems, traditional agriculture, far from being an irrelevant sideline, is a vital and potentially stabilizing force in an otherwise precarious scenario. But to what extent can native methods be buttressed, adapted or even reintroduced?

In the tropical lowlands, shifting cultivation needs assistance in order to enhance living conditions and protect the environment. Agricultural extension should ensure that appropriate, improved-yielding crop varieties are available and at an affordable price but this also depends on provision of fertilizer, pesticides and even certain low-energy tools. Diversification is also an obvious direction in which to go in order to avoid land exhaustion yet increase per capita output. In this connection one might advocate incorporating livestock into crop-grass and crop-legume alternations, planting arboreal fallows for commercial timber or going into fruit production. While it may be argued that pieces of original forest should be retained as they provide the only means by which forests can be reconstituted in future, there nevertheless does seem every reason ecologically and economically to encourage controlled and diversified tree cropping which in particular areas may help reduce erosion. Many tropical trees are candidates both for subsistence and commercial use. Cacao, oil palm, rubber and various timbers are examples of the latter while for subsistence purposes, the *ramón* or breadnut could be developed as has the breadfruit in Africa and Hawaii, and the pandanus nut and nitrogen-fixing casuarina in Papua-New Guinea. Indeed there is little doubt that the Mayas laid great emphasis on the produce of planted tree species. As far as agriculture is concerned, there is no doubt that the traditional system of creating a miniature forest of mixed cultivars offers a more varied diet, an increased period of harvest, better nutrient utilization and improved protection against pests and diseases. Nowhere on earth is it more important than in the tropical lowlands to create an artificially structured ecosystem for the optimum working of agriculture. Minimum or zero tillage has recently been recommended together with the use of crop and weed residues as a mulch for weed suppression and maintenance of soil fertility. Pre-Hispanic agricultural tools were very simple in design which further suggests that our temperate concept of freshly ploughed soil is alien to indigenous tropical systems and that intermediate technology may be more appropriate after all.

In the wetlands, intensive horticulture offers scope for community development and has attracted much interest in Mexico. Here, experimental chinampas have demonstrated their capacity for multiple cropping and in a matter of weeks, crops of manioc, rice, corn, beans, melons, tomatoes, alfalfa and many other vegetables have been raised. But publicly sponsored schemes must eventually be handed over to local people and allocated ideally to private family ownership for which they are well adapted. In Mexico, the *ejido colectivo* may be capable of providing a framework within which this can occur. Nevertheless we may wonder whether such developments will prove to have long-term viability in a market-orientated economy. In pre-Hispanic times we assume that such systems developed by centralized influence rather than through local initiatives. The success of rural enterprises certainly depends upon the existence of markets but ironically, as proximity to urban areas increases, so also do alternative forms of employment present themselves. In the Mexican lowlands the oil industry inevitably proves an attractive alternative to being a landless *chinampero*.

In the highlands a continuing problem has been the quality of nutrition as well as the quantity of food produced per unit area. The latter is brought about through a combination of high elevation, aridity and local salinity. The best prospects lie in diversification, employing mainly indigenous rather than exotic biota for reasons already discussed. On the grounds of environmental management one may therefore advocate the de-emphasizing of sheep and cattle in the Central Andean altiplano while a variety of measures including culling should be taken to improve the productivity of herds of llama and alpaca. Improved utilization and integration of water resources could do much to raise the aspirations and even the standard of life in the highlands albeit within the context of largely collective ventures. The Cusichaca project in the Urubamba valley of the Peruvian Andes, initiated by Dr Ann Kendall in conjunction with the Instituto Naçional de Cultura Peruana, has recently contributed to the re-establishment of Inca irrigation canals. While one hopes similar initiatives will be taken elsewhere there remain problems of land ownership and the collective will in isolated communities to develop markets for sale of produce surplus to local needs. In the Lake Titicaca basin, the reintroduction of nutritious aquatic plants known to be acceptable to the local population has been advocated while the rehabilitation of relict ridged fields growing a wide variety of salt-tolerant crops could bring a large area into production once again.

The object of external aid to indigenous agriculture should be to create a momentum of production which can be sustained. While the initial success of developments depends on a process of education and genuine involvement of local people, the long-term future depends on market opportunities being made available. Without this realization the process merely becomes one of giving with one hand and taking away with the other. The ultimate constraints against proposals for the envigoration and diversification – even reintroduction – of indigenous agriculture are then economic, social and political rather than physical or biological; in the words of Denevan (1970) 'there are no environmental limitations to the development of agriculture only cultural limitations'. It has not been the purpose of this essay to discuss in detail such constraints, which in any case vary substantially in type and degree from region to region. Yet failure to solve the agriculturally based problems of rural areas – by far the most widespread in the world today and reflected in increasing revolutionary activity – will only exacerbate the growing urban deprivation so manifest in the *barriadas* of major Latin American cities. It is here that the failure of successive governments to treat seriously and sincerely the needs of rural areas is most tragically manifest.

Further reading

Arnason T *et al.* (1982) 'Decline of soil fertility due to intensification of land use by shifting agriculturalists in Belize, Central America', *Agro-Ecosystems* **8**, 27–37. (Representative of range of papers dealing with contemporary stability of traditional lowland systems.)

Bronson B (1978) Angkor, Anuradhapura, Prambaran, Tikal: Maya subsistence in an Asian perspective, in **Harrison P D** and **Turner B L** (ed.) *Pre-Hispanic Maya Agriculture*, Ch. 14, pp. 255–300 University of New Mexico Press, Albuquerque.

Browman D L (1981) 'Prehistoric nutrition and medicine in the Lake Titicaca basin', in **Bastien J W** and **Donahue J N** *Health in the Andes*. American Anthropological Association. (An ethnobotanical summary with suggestions for future resource development.)

Brush S B, **Carney H J** and **Huaman Z** (1981) 'Dynamics of Andean potato agriculture', *Economic Botany* **35**, 70–88. (Modern pressures threatening the inter-regional stability of indigenous potato culture.)

Clayton E (1984) *Agriculture, Poverty and Freedom in Developing Countries*. Macmillan, London. (Up-to-date assessment of political and economic realities connected with survival and development of traditional methods.)

Darch J P (ed.) (1983) *Drained Field Agriculture in Central and South America*. B.A.R. International Series 189, Oxford, 263 pp. (Comprehensive set of papers by most of the contemporary researchers on pre-Hispanic ridged and drained field systems.)

Denevan W M (1970) 'Aboriginal drained field cultivation in the Americas' *Science* **169**, 647–54. (Pioneer research into pre-Hispanic agriculture. The evidence of field systems.)

Denevan W M and **Mathewson K** (1983) 'Preliminary results of the Samborodón raised field project, Guayas basin, Ecuador', in **Darch J P** (ed.) *Drained Field Agriculture in Central and South America*, Ch 9, pp. 167–181. B.A.R. International Series 189, Oxford.

Donkin R A (1979) *Agricultural Terracing in the Aboriginal New World*. Viking Fund Publications in Anthropology No. 56, University of Arizona Press, Tucson. (A scholarly and indispensable source on terrace agriculture, its morphology and regional variations extending from the United States to Chile.)

Farrington I S (1974) 'Irrigation and settlement pattern: preliminary research results from the north coast of Peru', in **Downing T E** and **Gibson McG** (eds.), *Irrigation's Impact on Society*. University of Arizona Press, Tucson, Anthropol. Papers No. 25.

Farrington I S (1980) 'The archaeology of irrigation canals, with special reference to Peru', *World Archaeology* **11**, 287–305. (Assessment of role and hydraulics of irrigation canals.)

Flannery K V (ed.) (1982) *Maya Subsistence: Studies in Memory of Dennis E Puleston.* Academic Press, New York. (A highly regarded documentation of regional pre-Hispanic Maya technology and ecology together with historical and modern derivatives.)

Flenley J R (1979) *The Equatorial Rainforest: a Geological History.* Butterworth, London. (A basis for understanding recently changed views on the environmental history of the equatorial belt.)

Gade D W (1975) 'Plants, man and the land in the Vilcanota Valley of Peru', *Biogeographica*, **6** Dr W Junk, The Hague. (Inventory of plants used by man and comments on agricultural practices.)

Greenland D J (1975) 'Bringing the green revolution to the shifting cultivator', *Science* **190**, 841–844. (The appropriateness of the introduction of new varieties and techniques to the small scale agriculturalist.)

Gross D R *et al.* (1979) 'Ecology and acculturation among native peoples of Brazil', *Science* **206**, 1043–1050. (The effects of technological change on a forest community.)

Gudeman S (1978) *The Demise of a Rural Economy: from Subsistence to Capitalism in a Latin American Village.* Routledge and Kegan Paul, London. (Ecological and social changes arising through changes from subsistence to a cash crop economy.)

Harrison P D and **Turner B L II** (eds.) (1978) *Pre-Hispanic Maya Agriculture.* University of New Mexico Press, Albuquerque. (A major collection of case studies and review articles on the archaeology and ecology of early agriculture in Mesoamerica.)

Herold L C (1965) *Trincheras and physical environment along the Rio Garilan,* Chihuahua, Mexico. University of Denver, Dept. of Geography Research Paper No 65–1.

Johnson A (1983) 'Machiguenga gardens' in **Hames R B** and **Vickers W T** (eds.) *Adaptive Responses of Native Amazonians,* Ch 2, pp 29–63. Academic Press, New York.

Keeley H (1985) 'Soils of prehispanic terrace systems in the Cusichaca Valley, Peru', in (ed.) **Farrington, I S** *Prehistoric Intensive Agriculture in the Tropics* BAR International Series, 232, Oxford.

Kirkby A V T (1973) 'The use of land and water resources in the past and present Valley of Oaxaca, Mexico', *Memoirs of Museum of Anthropology,* University of Michigan, No. 5.

Parsons J R (1976) 'The role of Chinampa agriculture in the food supply of Aztec Tenochtitlán,' **Cleland C E** (ed.) *Cultural Change* *and Continuity*, Ch. 12, pp. 233–257. Academic Press, New York. (Investigation of the networks of tribute organized to support the Aztec capital.)

Pickersgill B and **Heiser C B Jnr** (1978) 'Origins and distribution of plants domesticated in the New World tropics', in **Browman D L** (ed.) *Advances in Andean Archaeology*, pp. 133–165. Mouton, The Hague. (An obvious starting point for any appreciation of indigenous American ethno-botany. An eminently readable article in a useful collection on the Andes.)

Pickersgill B and **Smith R T** (1981) 'Adaptation to a desert coast: subsistence changes through time in coastal Peru', in **Brothwell D** and **Dimbleby G W** (eds.) *Environment Aspects of Coasts and Islands*. British Archaeological Reports: International Series No. 94, pp.89–115. Oxford. (Evolution of subsistence patterns and relationships between coast and Andean interior.)

Rindos D (1983) *The Origins of Agriculture*. Academic Press, New York. (A review of ideas which conveys mostly the current tenor of materialistic thought on this vexed and inevitably esoteric subject.)

Sanders W T and **Price B J** (1968) *Mesomerica: the Evolution of a Civilization*. Random House, New York. (Excellent, readable account of mesoamerican history if now somewhat dated.)

Siemens A H (1983) 'Wetland agriculture in pre-Hispanic mesoamerica', *Geographical Review* **73**, 166–181. (Reveals the scale, cultural cohesion and ecology of raised bed agriculture in pre-Hispanic times.)

Smith R T (1979) 'The development and role of sunken field agriculture on the Peruvian coast'. *Geographical Journal* **145**, 387–400. (A little known pre-Hispanic adaptation to the particular hydraulics of the arid Pacific coast.)

Smith R T (1984) *Environment and Indigenous Agriculture in the American Tropics*. Working Paper 405, School of Geography, University of Leeds. (A fully bibliographic treatment of the subject matter of this chapter.)

Turner B L II and **Harrison P D** (1981) 'Prehistoric raised-field agriculture in the Maya lowlands', *Science* 213: 399–405.

Weaver M P (1981 2nd edn.) *The Aztecs, Maya and their Predecessors*. Academic Press, New York. (Lavishly illustrated overview of significant aspects of mesoamerican archaeology including agriculture and the use of resources.)

CHAPTER 4

Early industrial patterns

John Dickenson

It can be argued Latin America remains a 'non-industrialized' area, in that in 1981 only in Argentina, Uruguay, Peru, Brazil and Nicaragua did manufacturing contribute at least 25 per cent (the average for the World Bank's 'industrial market economies') of gross domestic product. No Latin American country in 1980 had one-third of its labour force employed in secondary activities. Yet if the continent is still industrially underdeveloped, the pattern and structure of such industrial activity as there is has evolved over several centuries, from a few small scale artisan workshops to a much more complex mixture of basic industries, transnational corporations and surviving traditional industries.

The earliest industrial activity pre-dates the arrival of Columbus, since some Amerindian groups had established considerable craft traditions in wood and metalworking, pottery and textiles. Significant development of industry, however, post-dates Columbus and Cabral, and the nature of Spanish and Portuguese rule has been significant in shaping the process and pattern of industrial activity in Latin America until recently. The colonies of the New World were seen as lands to be exploited, to provide primary produce and wealth for the metropolitan countries and markets for their goods. Such a pattern persisted after the achievement of independence, when other countries succeeded Spain and Portugal as the dominating trading partners.

Colonial beginnings

The basic pattern of industry which evolved in the colonial period was one in which manufacturing activities were mainly confined to the preliminary processing of primary products such as agricultural commodities and minerals for export, and to the production of the most basic goods needed to provide sustenance for the local population, namely food, clothing and shelter. The consequence of this economic structure was that these activities tended to develop in

70

a limited range of locations – at the point of production of export commodities, at the port of export, or in the principal cities and markets. These functions sometimes overlapped, but included foci such as the mining areas of Mexico, Upper Peru and Minas Gerais, the sugar zone of north-east Brazil, the vineyards of Peru and western Argentina, ports such as Vera Cruz, Puerto Bello and Salvador, and areas of indigenous population clusters such as the former Aztec and Inca territories in highland Mexico and the central Andes.

The peaks of the mining booms ended in Spanish America by the early seventeenth century and in Brazil by the mid-eighteenth, so that agriculture and pastoralism were the principal economic activities of the late colonial period. However, three centuries of Iberian rule established economic patterns and processes which remained significant after independence. Latin America had become a source of distinctive raw materials for the metropolitan powers, and its economy was shaped in their interest, with the discouragement of autonomous economic activity. The basic pattern of settlement was firmly established. Most of the major nodes of population and economic activity were established before 1600, around Mexico City, Santo Domingo, Bogotá, Quito, Lima-Cuzco, Santiago, Asunción, São Paulo, Salvador and Recife. Many of the lineaments of Latin America's transport system were also established before 1800, both in terms of the ports serving trade to the North Atlantic, and overland routeways. Such links had significant impact and legacy on the pattern of settlement and economic activity. There were also areas neglected in the colonial period, particularly the arid and semi-arid areas of northern Mexico, the Atacama desert, the Gran Chaco, Patagonia and Brazilian sertão, and the rain forests of Amazonia. In some of these areas also, the indigenous population resisted Iberian advance.

Most of the colonies of Latin America secured their political independence by 1830. Though Portuguese America retained its unity, Spanish America fragmented into a number of states, of varying size and colonial experience, with differing societal structures and economic patterns. There were also internal differences – between city and countryside, latifundista and peasant, Creole élite and mestizo proletariat, Amerindian and African, and various regional interests. Although political independence had been secured, the basis for independent economic development, much less industrialization, was limited.

The nineteenth century: independent and dependent industrialization

In fact the economic control exercised by Spain and Portugal had been loosened before independence, so that the colonies had developed (or legitimized) wider trading links. However, the local élites saw no need to alter the export-oriented economy which had evolved, or to forego the imported goods to which they had become accustomed. In consequence a pattern of external dependency continued, and was progressively heightened by the industrializing countries of western Europe, even more anxious than Iberia to secure raw materials and markets for their developing economies.

Initial change was slow, for the resources Latin America had to offer were limited, such as precious metals, hides, skins and tallow, or else had experienced price declines, such as sugar and cotton. The full impact of the new trading relationships did not emerge until after about 1840. Elements of colonial industrial activity persisted in the limited processing of such export goods, and in the craft production of goods to satisfy basic needs, together with limited government sponsored shipyards, mints and arsenals and similar activities. There were some industrial initiatives in the early nineteenth century. In Mexico, for example, craft traditions provided the basis for cotton textile production on a factory scale in Puebla, Queretaro, Guadalajara and other centres. In 1843 there were fifty-seven mills in Mexico, but problems of transport, access to raw materials and opposition to the industry's tariff protection resulted in their decline by mid-century. In Brazil, even before independence the Portuguese prince regent had given encouragement to industrial development, inviting European technicians to help establish ironworking, shipbuilding and other industries. Paradoxically, Brazilian ports had been opened to non-Portuguese ships and favourable tariff conditions conceded to the British at the same time, such that the impact of the latter inhibited the former initiatives, and the projects lapsed.

Before 1840 the new nation-states were establishing themselves, few had commodities to trade, and external demand from Europe was still relatively low. Such trading links as were established tended to focus on the major ports, perpetuating the pattern which had been established in the colonial period.

The development of Latin America in the nineteenth century is closely linked to the 'take-off' of this trading relationship, essentially a product of the then current notions of free trade. The evolution of a world economic system and of an international division of labour heightened Latin America's role as supplier of primary produce to the markets of Britain and later Europe and North America, and a recipient of their manufactured goods. Such a relationship was a stimulus, as in the colonial period, to export-commodity processing activities and to some expansion of essential domestic industries; but it also provided some basis for the gradual evolution of more substantial industrialization.

The economic advance of the countries of the North Atlantic during the latter part of the nineteenth century generated not only a demand for raw material inputs and the need for markets for output, but the means by which these transactions could be carried out, with the development of steamships, and an increase in the scale, price and speed at which goods could be transported. Between 1840 and 1860 the world merchant fleet doubled in size, and by 1913 it had increased almost seven-fold. The development of the railway had similar consequences for land transportation, opening up and integrating new raw material sources and markets.

Commodity trading and transport

The ruling élites of Latin America espoused these trends and initiatives, seeing the role of their countries as supplying the agricultural and mineral raw materials they could produce cheaply, in return for manufactured goods they were unable to produce. Many saw this relationship as being the means to introduce European and North American 'progress' into their territories and, of course, to provide themselves with wealth from the sale of sugar, meat, coffee, minerals and other products. Before 1875 Britain tended to dominate Latin America's trading structures, but during the late nineteenth century this situation changed with increasing participation from the USA, Germany and other European countries.

The commodities sought by the neo-colonial powers took three major forms. The first group consisted of temperate agricultural produce, primarily to meet the expanding food needs of the urbanizing-industrializing northern nations. Demand for meat and grain had the consequence of extending the frontier of settlement and exploitation especially in Argentina and Uruguay, and was closely associated with the introduction of new modes of transport and farming techniques. The second commodity group was that of tropical crops, such as sugar, coffee, cacao, bananas and tobacco, frequently creating areas of large scale monoculture and generating new patterns of settlement and transport. The other commodity group sought by the developed world was minerals – not the precious metals exploited by the Iberians but more mundane materials for northern fields and factories – guano, nitrates, copper, tin and later, iron and petroleum.

These commodities became major elements in the trading pattern and development of Latin America, profoundly influencing the scale and nature of economic development in the nineteenth and early twentieth century. A crucial element in their exploitation was the improvement in transport. In the early post-colonial period its deficiencies tended to perpetuate the pattern of concentration of economic activity close to the coast, and in the principal cities such as Rio de Janeiro, Buenos Aires, Santiago and Lima. The introduction of the railway, however, brought profound change, facilitating the opening of interior areas and resources.

The first significant railway development was in Cuba, in 1837, where it was introduced to facilitate the transport of sugar and coffee. However, more general development began only after 1850, with some of the earliest ventures being short lines running inland

Plate 5 A reconstructed blast furnace at Itabirito, Minas Gerais. The furnace was originally built in 1888 and was a pioneer in the modern phase of Brazil's iron and steel industry. (Photo: J. P. Dickenson)

Plate 6 The Cia. Belgo-Mineira steelworks at Monlevade, Brazil. It was opened in 1937 with investment from the Belgo-Luxembourg ARBED company, to utilize local iron ore and charcoal.
(Photo: J. P. Dickenson)

from the major cities. In many parts of Latin America considerable technical problems in railway construction arose in overcoming such obstacles as the Andes, the Serra do Mar in Brazil and in the Sierra Madre in Mexico. The completion of railways such as the Santos–São Paulo (1868), Vera Cruz–Mexico City (1873) and Mollendo–Puno (1877) were of major significance for resource exploitation and development.

As Fig. 4.1 indicates, the most substantial railway systems were built in Argentina, Brazil and Mexico, which in 1940 accounted for 75 per cent of the region's network. It is also evident that, in most cases, the bulk of the railways were built between 1880 and 1920, with only Colombia, Honduras, El Salvador, Ecuador and Bolivia having significant extensions to their railway networks after 1920. The bulk of these networks were built for exploitative purposes, intended to facilitate the removal of primary produce to the ports and abroad, rather than provide co-ordinated internal transport systems for or between the various countries. Most of the systems are therefore simple, consisting of lines running from the coast inland. This is the case, for example, for the railways of north-east Brazil, Peru, Ecuador, Venezuela and Central America. Only Mexico, Cuba, Uruguay, pampean Argentina and south-east Brazil have denser and more complex nets, and few international links were built. These differing patterns reflect, at least in part, contrasts in the nature of the export economy, with single lines providing access to mines, and more complex nets serving the more extensive production of grains, coffee, cattle and other agricultural commodities. In both cases, however, the rail system tended to heighten the dominance of the coastal towns.

The railways were also significant in the nineteenth-century development of Latin America, in being closely linked to foreign investment in the continent. Though governments encouraged railway building, often with generous concessions and guaranteed profits, much of the investment and technology was foreign, particularly from Britain and the USA. In 1914 over four-tenths of British investment in South America was in railways, controlling, for example, the bulk of the Argentinian and Uruguayan systems.

The railways, in a sense, confirmed and strengthened the spatial pattern of development begun in the colonial period, paralleling its routeways, increasing the significance of existing coastal settlements, and serving the interests of the economies of the North Atlantic. The search for, and exploitation of, new resources did open up some new areas and extend the area integrated into the international economy. Foreign capital was also significant in other sectors, frequently in its own interests, in improving ports and shipping, providing urban services such as transport and electricity, and in direct investment in mining and some industrial development. Such developments facilitated Latin America's role as a supplier of primary products, so that by 1914 the continent was a major source of sugar, cereals, coffee, cacao, livestock products, rubber, fertilizers, and

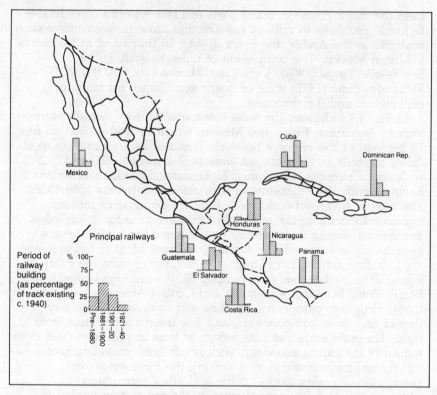

Fig. 4.1 The evolution of the railway system to 1940 (*Source:* from Mitchell B R (1983) *International Historical Statistics: the Americas and Australasia.* Macmillan, London)

minerals such as tin and copper. The detailed spatial pattern of these activities and their contribution to incipient industrialization varied.

Temperate agriculture produce

Of the three major groups of export commodities, temperate and tropical agricultural produce initially made limited contributions to industrialization, but in some cases were significant stimuli to factory production. A striking example of this process and pattern comes from the pastoral industries of the pampean grasslands of Argentina, Uruguay and southern Brazil. Feral cattle had provided the basis for a very extensive pastoral industry in the colonial period, with cattle being slaughtered for their hides, which were crudely tanned. The early nineteenth century saw more ordered cattle raising and meat processing with the development of the *saladeros* (meat salting factory). The earliest developed at Sacramento on the Uruguayan coast and then along the north shore of the Plate estuary, spreading to Argentina in the early nineteenth century. This was essentially a

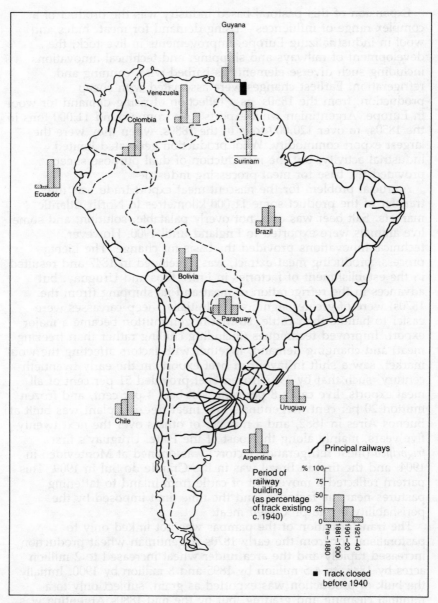

Fig. 4.1 (contd.)

crude, seasonal activity of salting and drying beef to provide low grade meat for the slave populations of Brazil and the Caribbean. The rudimentary industry also yielded hides, grease, tallow, bone meal and gelatin, and expanded rapidly. Argentinian output rose from 60,000 cattle in 1822 to 350,000 in 1827. The *saladeros* were located along the coast of the Plate and in Rio Grande do Sul.

Expansion of this pastoral based industry was the product of a complex range of influences – rising demand for meat, hides and wool in industrializing Europe, improvements in livestock, the development of railways and shipping, and technical innovations including such diverse elements as barbed wire, canning and refrigeration. Earliest changes were associated with sheep production, from the 1850s, as a reflection of rising demand for wool in Europe. Argentinian wool exports rose from around 11,000 tons in the 1850s, to over 120,000 tons in the 1880s, when they were the largest export commodity. Wool production generated limited industrial activity, but the introduction of dual purpose sheep provided the base for meat-processing industries.

A crucial problem for the nascent meat export trade was the transfer of the product some 11,000 kilometres to North Atlantic markets. Salt beef was one, not overly palatable, solution, and some live animals were exported to England until 1900. However, technical innovations provided the basis for change. The Liebig process, producing meat extract, was developed in 1847 and resulted in the establishment of factories in Entre Rios and Uruguay, but advances in the refrigeration of meat and its shipping (from the 1870s) were of greater significance. Initially sheep carcasses were easier to handle than cattle, so that frozen mutton became a major export. Improved techniques (including chilling rather than freezing meat) and changing demand, together with factors affecting the wool market, saw a shift in favour of beef exports in the early twentieth century, such that by 1907 frozen beef provided 51 per cent of all meat exports, live cattle 8 per cent, salt beef 4 per cent, and frozen mutton 20 per cent. Argentina's first meat freezing plant was built at Buenos Aires in 1882, and a number of others over the next twenty-five years, mainly along the coast of the Plate. Uruguay's first *frigorifico* (meat refrigeration factory) was opened at Montevideo in 1904, and the first in Brazil was in Rio Grande do Sul in 1909. This pattern reflected a movement of cattle from inland to fattening pastures near the factories, and the time limit imposed by the perishability of the processed meat.

The transformation of the pampas was not linked only to pastoralism, for from the early 1870s Argentinian wheat production increased rapidly and the area under wheat increased to 2 million acres by 1888, to 4.5 million by 1895 and 8 million by 1900. Initially the bulk of production was exported as grain, subject only to a minimal cleaning and grading, but by the mid-1880s Argentina was providing most of its own flour and had begun to export, but during the nineteenth century locally processed wheat made only a small contribution to exports. In 1900 flour provided less than 3 per cent of the country's wheat and flour exports. The rise of a flour milling industry was therefore more linked to domestic demand. In 1889 two-thirds of the country's steam-powered mills were located in Santa Fe province, the largest wheat producer, but increasingly the industry shifted to the principal wheat export port and major domestic market, Buenos Aires. By 1914, 55 per cent of flour output was in the capital and its province.

The broad economic and specifically industrial consequences of this 'agricultural revolution' on the pampas were complex and to an extent contradictory. Rising demand from Europe generated increased production of grain, sheep and cattle. This, combined with the positive effects of railway building, extended the area in agro-pastoral use and generated a distinctive pattern of land use. The export orientation of production and of the nature of the transport net focussed on the coast, and particularly Buenos Aires, so that the processing of these commodities, whether for export or domestic use, tended to concentrate there. This emerging primacy was not only associated with the produce of the pampas for, though the initial processing of crops of sugar and grapes took place in the producing areas of the north-west, the principal sugar refinery was in Rosario, and Buenos Aires had a significant role in the blending and bottling of the wines of Mendoza and San Juan.

Tropical agricultural produce

Elsewhere in the continent tropical export crops also provided some basis for processing activities, though this tended to be limited. In the case of sugar, for example, though there was some technical progress and increase in the size of holdings, processing tended to be carried on in relatively small scale sugar mills or *engenios*. In Cuba there were some 2,000 such mills in 1860, and in north-east Brazil such plantation-located mills, using slave labour and producing crude sugar were dominant until late in the century. After 1875 larger units of production, the *usinas*, began to emerge. They demanded larger inputs of raw material and larger investment, and resulted in the amalgamation of holdings, controlled by companies rather than individual plantation owners. Even so, it was not until the twentieth century that they became the leading source of sugar. In Cuba the large modern mills (*centrales*) began to develop after 1900, particularly in new cane lands in the east of the island. These tended to be foreign controlled and associated with very large land holdings, in contrast to the smaller Cuban mills and their associated plantations in the older cane areas. Between 1902 and 1927 over sixty such *centrales* were built, principally with North American capital, located on the plantations and shipping out the sugar via the dense rail net and a series of specialized ports such as Nuevitas, Santiago and Antilla. Similar processes occurred in the sugar lands of Peru, Mexico and the Caribbean.

One of the continent's principal tropical export commodities was coffee, and this had, in some areas at least, significant impact on the direct and indirect process of industrialization. In Colombia the early post-independence period was one of economic stagnation and decline, and export production of tobacco and quinine provided little basis for industrial activity. The rise of coffee however, provided a more positive stimulus. Production rose from about 6,000 tons in

1880 to 66,000 tons in 1915, and increased its contribution of export earnings from 12 per cent to over 50 per cent. Expansion was concentrated in the Antioquia region and formed part of a small-farm economy. Coffee processing requires only simple equipment – depulping and drying machinery, and bags for the dried beans. This provided the basis for simple industries producing machinery and sacking, the latter initially as an artisan industry and later as an urban factory industry. In addition the small-farm basis of production meant that the farmers were less prone to the 'conspicuous consumption' of imported goods characteristic of some nineteenth century élites in Latin America. Instead their coffee income provided demand for local textiles and other necessities. This encouraged the development of craft textile activities, leather working, pottery and iron founding. In turn, rising prosperity and the development of an expanding railway system generated the basis for factory scale industries such as brewing, sugar refining and sweet manufacture, and textiles, all established in the first decade of the twentieth century, with Medellín as a major focus.

A very similar process, though on a larger scale, took place in south-east Brazil, especially in São Paulo state where the rise of the coffee economy from the 1850s had major impact. Coffee generated capital from export earnings, land sales and wages, immigration (1.2 million immigrants between 1896 and 1906) provided a growing free market and some industrial skills, and there was major infrastructural development of railways, and of cheap hydroelectric power from 1899. These factors provided the basis for early industrialization, with metalworking and textiles for the coffee industry and other agricultural products, and subsistence food, clothing and building material production for local use. The pioneer industries were those with high weight-to-cost ratios able to compete with imported manufactures, and which used local raw materials or imports which gained bulk during processing. Textile mills began to develop at water power sites from the 1870s. These used local cotton, cultivation of which had increased during the American Civil War, but which lost its export market with the end of hostilities. Three cotton mills were established in São Paulo state in the 1870s and a further four between 1880 and 1884.

São Paulo's railway system was also a key element in the state's industrialization for not only did it facilitate the rapid advance of the coffee frontier and the rise of the export economy, but railway workshops provided centres of metal working activity and fostered engineering skills. In addition the railway lines, junction towns and railheads provided nodes of urbanization and foci for early industrial activity. However the focussing of the rail net on the town of São Paulo (which increased in size from 65,000 to 579,000 inhabitants between 1890 and 1920) and thence to the coffee port of Santos, made the former the principal centre of industrial growth. By 1895 the city had 121 mechanized factories, though these were mostly relatively small in size. Only eleven of these plants had more than one hundred workers – in the textile, clothing, match and iron founding industries, and in railway workshops.

In addition to these examples, other areas of agricultural export development in the nineteenth century included coffee in Costa Rica, Guatemala and Mexico, sugar in Peru and the Guianas, and rubber in Amazonia. In such cases the export product provided some basis for local processing activities, but not necessarily the impetus to more substantial, independent industrial development.

The broad tendency of the rise of the agricultural export sector, both tropical and temperate, in Latin America, was to extend the territory involved in commercial activity, by opening up new land, or intensifying the use of older settled areas. This generated some industrial activity in these agricultural areas, generally of a fairly rudimentary nature, but it also tended to foster more substantial, diverse and sophisticated industries in the ports, whether concerned with processing for export, the processing of imports, or other activities concerned with the domestic market. Such nodes were generally the principal urban centres of the various countries.

Minerals

If the agricultural export sector tended to generate some degree of diffusion in the economic activities of nineteenth-century Latin America, the third element, of mineral exports, fostered a pattern of concentration. This was inevitable, given the normal pattern of distribution of mineral resources. In contrast to the colonial experience the new mineral exports were more mundane in character, providing inputs for the fields and factories of industrializing Europe and North America, rather than gemstones and precious metals. Among the earliest of these developments was the export of guano as a fertilizer, from islands off the Peruvian coast. This required no processing, so that the direct impact of this activity from 1840–1880 in generating industry was negligible. However output, which averaged over 400,000 tons a year between 1862 and 1873, provided a large share of federal income, and thus capital for investment in mainland agriculture, roads and railways.

The guano boom was diminished by exhaustion of the resource, and competition from the nitrates of the Tarapacá and Antofagasta areas of the Atacama desert. These resources had been exploited from the 1830s but became of greater significance when their use in the manufacture of explosives, as well as fertilizer, was recognized. After the War of the Pacific (1879–83), exploitation of the nitrate deposits took place in what had become northern Chile. This generated a classic and ephemeral exploitative economy, in which the nitrates were mined, refined in *oficinas* and moved by rail to a series of ports. In an essentially hostile environment this was an alien activity. It remained viable so long as resources remained and there was adequate demand for the product. While these conditions prevailed the boom nitrate towns and ports and their immigrant populations could be sustained with foodstuffs, supplies and even water brought in from elsewhere. At its peak during World War I, there were some 65,000 workers in the industry and a regional population of over a quarter of a million. However, competition from synthetic nitrates destroyed Chile's monopoly and by the 1930s the importance of the industry had declined and many of the oficinas had closed.

In its heyday, the Chilean nitrate production represented a significant concentration of industrial activity, albeit of a rather rudimentary kind. Once its basis had been undercut it declined, with few spin-offs into other activities. Its prosperity had, however, provided some stimulus to developments elsewhere in Chile, since taxation on nitrate exports had provided over half of government revenue between 1890 and 1920. The industry was also characterized by its association with foreign capital. In 1901 over 80 per cent of the investment in the industry was in foreign hands (British, German and Spanish); only 15 per cent was Chilean.

Similar features characterized other mineral-based activities in the nineteenth century – foreign capital, spatial concentration, export orientation, subservient transport systems and limited industrial spin-off. Northern Chile had also been a source of copper exports in the mid-nineteenth century, with copper being pre-eminent in the export economy in this period, until overshadowed by nitrates, competition from other producers, and exhaustion of the richest sources. In this latter respect the rudimentary nature of Chilean production methods could not compete with those of American and Iberian sources. Production tended to be small scale, with limited mechanization and dependence on traditional energy sources, such as wood, rather than coal and steam engines. In consequence, as in the case of nitrates, copper production in this first phase was unable to cope with external influences and copper mining and its associated activities went into decline from the 1890s.

In Mexico there was a legacy of colonial mining activity; the silver mines were modernized using British capital and new techniques from mid-century, and new ore-milling and refining techniques were introduced from the USA after 1875. The core of this mining and processing activity was in the established colonial areas of Pachuca, Guanajuato, Zacatecas and Parral. In 1880 gold and silver remained Mexico's principal mineral exports, but by 1900 lead, zinc and copper had become more important, fostered by considerable North American investment in mines and smelters. In addition, towards the end of the century, Mexico was one of the pioneers in coal production, developing mines in Coahuila to provide fuel for railways and later iron and steel production, significant in a continent seemingly deficient in solid fuels and, at that time, heavily dependent on coal imported mainly from Britain.

Growth in industrialization

These elements then were important in the limited nineteenth century industrial development of the continent – an orientation towards export production of agricultural goods and minerals, a considerable element of foreign investment, both in directly productive activities and in supporting infrastructure, and a tendency to spatial concentration of this incipient industrial activity, in the area of production and the exporting ports (which were frequently also the major cities). Although this export-led industrialization was of great significance, there was also some domestically orientated development of essential commodities, as the export sector sustained prosperity and population growth towards the end of the century. The general trend of these two strands of precocious industrialization was towards concentration in a few centres.

In addition to the dependence on external demand and priorities towards industrialization in this period, there were a variety of other checks. A fundamental one was the nature of the neo-colonial economy, in which Latin American primary products were traded for European and North American manufactures. It was not easy for the infant industries of the continent to compete with the products of a highly dynamic industrial revolution.

In the early post-colonial period political unrest inhibited the development process and served to deter foreign investors. Increased stability made the continent more attractive to such investment, both in productive activities and in related infrastructure. However, though foreign capital sought resources required by the industrial countries, the continent's resources remained imperfectly known and, particularly in the case of solid fuel, a perceived essential base of industrialization, Latin America appeared poorly endowed. The market for industrial goods was limited. A small élite sought, and could afford, the luxury of imported European manufactured goods, while even limited production of artisan, and craft industries, was affected by competition from imports. Significant segments of the population remained outside the market economy, because of their poverty, as members of Amerindian communities or as slaves. Total populations and therefore potential sources of labour and markets were small. Of the countries for which data are available, only Mexico and Brazil had more than 5 million inhabitants in 1880, and Argentina was the only other nation to reach this level by 1900.

Many Latin Americans and their governments therefore accepted the role as suppliers of raw materials in return for imported manufactures. There was also a surviving notion that manual labour and trade were activities inferior to those of the professions or land holding, and that investment in land was more worthwhile than investment in machines. Such factors may have served to inhibit industrialization, but the situation was not wholly negative. In addition to the export-linked processing industries there were direct and indirect stimuli to independent industrial growth. Export earnings did provide capital for investment, infrastructure widened and integrated the area of developed land, and provided a base for new economic activities. Population began to grow, both by natural increase as a consequence of medical advances, and, in some areas, by immigration. In some countries, such as Brazil, Argentina, Uruguay and Chile, immigrants appear to have played a significant role in industrialization. They contributed to the general process of economic expansion and to urbanization. They provided entrepreneurial and technical skills as well as contributing to the formation of an incipient urban working class. They formed an expanded market for industrial goods and generated a new demand for familiar European products such as wheat flour, pasta, wine and beer. Their savings, invested in branches of banks from their home countries, provided some industrial capital. Although most of these immigrants had been recruited as agricultural labourers or colonists, some migrated direct to the cities, and others soon moved there from the countryside. In consequence by 1920 almost two-thirds of the inhabitants of São Paulo city were foreign born or the children of foreigners: in Buenos Aires the figure was around 50 per cent. In 1915 there were three times as many foreigners as Argentinians in the country's manufacturing sector. Immigrants tended to be associated with more modern industries, such as chemicals and metal-working, whilst the local population was more commonly in more traditional activities, such as clothing, with less complex technology and organization. Substantial immigration to countries such as Brazil and Argentina was a response to the labour demands of the export economy, but served as an important stimulus to economic diversification and to industrialization.

Although the primary product export lobby continued to be strong there were innovators and entrepreneurs who sought to foster manufacturing. In the third quarter of the nineteenth century in Brazil, Viscount Maua was involved in encouraging not only developments in finance, transport and energy, but in the shipbuilding, iron and leather industries. People of immigrant origin were also early entrepreneurs in Brazil, particularly in São Paulo and Rio Grande do Sul. In some cases they provided the nucleus for important contemporary industrial organization. The evolution of the Matarazzo group provides a striking example, with evolution from a rural store into grain, pig and lard dealing and thence to the canning of lard and flour milling. The latter led to textile manufacture (for sacks), and cotton seed oil to soap manufacture, which in turn required packing cases, and thus a wood-working industry. Such evolution points to the rising production and demand for commodities for local use – food, cloth and packaging. Similar elements may be noted in Argentina where small factories sprang up to produce local necessities – clothing, footwear, drink, cigarettes, bread, pasta, building materials and furniture and in Mexico immigrants were also responsible for water-powered cotton and woollen mills around Orizaba from mid-century. The development of hydroelectric power in the 1890s provided the energy source for textile mills in Orizaba, Jalapa, Puebla and Mexico City. Expansion of large scale cotton cultivation in northern Mexico and the evolving rail net provided both raw material and access to wider markets.

There is certainly evidence of incipient industrialization in Latin America by the end of the nineteenth century, not only in the necessary processing of export commodities, but in production for domestic demand. The continent had begun to benefit from the diffusion of European and North American innovations – the railway, telegraph, electricity, medical advances, and similar developments. The internal distribution of these benefits was, however, uneven. Some areas and some segments of society remained beyond their impact. Conversely, rising urbanization and the emergence of an urban élite and middle class generated some pressure for modernization and change. Economic and cultural contact with Europe and North America prompted a desire for continuing progress, including industrial development. However, neither this enlightenment nor the wishes of the limited band of industrialists was able to generate profound change, given the continuing dominance of the agricultural élite in Latin America's political structure. There were few concessions to industrial interests in the way of tariff protection, access to credit, or direct involvement by government in manufacturing.

The pattern of industrialization at the end of the nineteenth century therefore was one in which a few countries had made some progress – Argentina, Brazil and Mexico, and to a lesser extent Chile, Colombia and Peru. This industry was based partly on export-oriented processing, and partly on production for limited domestic demand. In the latter case it was concerned with producing essentials, such as food, clothing and building materials. The scale of production tended to be small, with a significant artisan sector, and dependent on local raw materials and serving essential local markets. The spatial pattern of activity therefore involved location at the point of production or the ports for agricultural and mineral exports, while industries concerned with the domestic market tended to locate at the principal concentrations of populations (Fig. 4.2). This in part served to reinforce the locational pattern of the export sector, but there was also some dispersal, particularly of artisan industry, in smaller centres. The basic pattern, at both an international and intranational scale was already being clarified, with the three largest countries, Brazil, Argentina and Mexico having made most progress, and south-east Brazil, the pampas and the valley of Mexico emerging as the foci of economic development and industrial concentration, together with the principal cities of the other republics. Even so, the level of industrial development remained relatively low. In 1895 Argentina's second national census recorded 175,000 workers, and Brazil's first industrial census of 1907, 150,000 workers; Great Britain, in 1911, had over 8 million industrial workers.

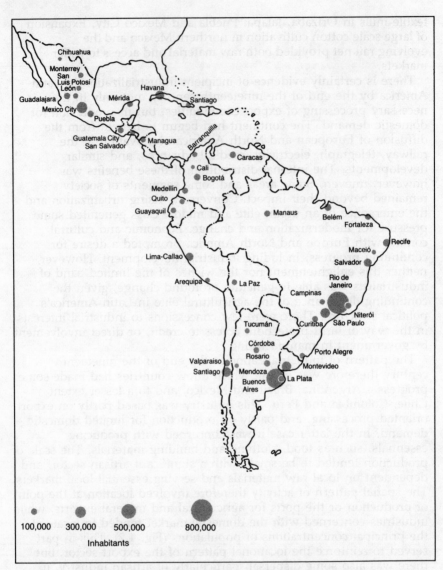

Fig. 4.2   Principal cities 1900 (*Source*: From Mitchell B R (1983) *International Historical Statistics: the Americas and Australasia*. Macmillan, London)

Progress in the early twentieth century

The expansion, diversification and spatial change which took place in Latin America's industrial sector in the first half of the twentieth century derived from a variety of stimuli. An important element was the identification of a wider and more substantial resource base.

Nineteenth-century economic theorists argued that cheap and abundant power was an essential element for industrialization, yet the continent was apparently deficient in the fundamental fuel, coal. In 1900 only Mexico and Chile had coal industries of note, though producing only about 350,000 tons each, against United States output in excess of 240,000,000 tons! In the period before 1940 production also began in Brazil, Colombia and Peru, but on a world scale Latin America remained insignificant. Such developments did provide some basis for industries using coal as a raw material or fuel, though the coal produced tended to be of poor quality.

Of greater significance was the utilization of alternative forms of energy sources. With the recognition of petroleum as a source of fuel and power, some Latin American nations proved to be well endowed. Mexico was the first significant producer, from 1901. Large scale production began in 1910 and between 1918 and 1927 Mexico was the world's second largest petroleum producer, after the USA, eventually losing its position to Venezuela. Commercial production began in Argentina in 1907, Venezuela in 1917 and Colombia in 1921, and by 1940 these countries together with Peru, Ecuador and Bolivia were significant oil producers. In some cases oil provided an important export commodity, and in others an alternative to deficient solid fuel resources. It was also the basis for some industrialization, with the development of oil refineries and of petroleum product industries. Such development provided the stimulus to new areas of economic activity and industrialization, in the Mexican Gulf Coast, around Lake Maracaibo, in western Ecuador, and in the Chubut and Mendoza areas of Argentina. The oil industry is also of significance in that it was amongst the first in which the State began to take a direct interest in the organization and activities of the productive sector. Although many of these resources were initially exploited by foreign companies which were to become the seven 'international major' oil firms (Esso, Mobil, Gulf, Chevron, Texaco, Shell & BP), several Latin American countries were pioneers in the expropriation of these interests, and the establishment of domestic companies to control the exploration for and production of oil. This action, by Argentina and Chile in 1927, Bolivia in 1937 and Mexico in 1938, reflected an emerging view that certain sectors of the economy were of such crucial importance that their development must be in the hands of government and not of foreign interests.

The continent's needs were also assisted by the recognition of its considerable water power potential, estimated at 13 per cent of the world total in the early 1940s. As a source of power, hydro-electricity has been significant in sustaining the location of some large energy-consuming industries, but more significantly in supplying the electricity needs of factories located in the major industrial centres. The structure of development in this sector before 1945 was varied, with some significant involvement of foreign capital, such as the major stimulus provided to the industrial growth of São Paulo and Rio de Janeiro by the American and Foreign Power Co. Inc. and the Canadian-owned Brazilian Traction, Light and Power Co. Ltd. By the 1940s these companies were providing 80 per cent of Brazil's electricity. Elsewhere, as in Chile, Colombia and Uruguay, government was active in encouraging hydroelectric projects, while in other cases individual concerns developed sites to meet their specific needs. Thus in the late 1930s Peru's largest h.e.p. plant was owned by the Cerro de Pasco company.

The continent's mineral resource endowment was also becoming more evident, both in terms of new sources of known minerals such as iron, manganese, copper and tin but also newly useful materials such as bauxite, nickel and ferro-alloys. Though their exploitation tended to perpetuate the nineteenth-century pattern of semi-processing prior to export, they also provided potential raw materials for domestic industry, as did the agricultural sector.

An important complex of political, economic and social factors also provided an impetus towards industrialization in the early twentieth century. Intellectual arguments favouring industrialization were strengthened by external influences from the world economy. Dependence on primary product exports and free trade became more vulnerable with fluctuations and declines in commodity prices, particularly affecting those countries heavily dependent on only one or two commodities. The impact of World War I has generally been argued to have been a positive stimulus to industrialization, cutting off Latin America from markets and reducing manufactured goods imports and thus prompting domestic manufacture. The weakening of ties of trade and investment is claimed to have facilitated autonomous industrialization and growth, particularly in Argentina, Brazil, Chile and Mexico. In recent years this argument has been questioned, with evidence to suggest that though output, using spare capacity, increased to meet some scarcities, the amount of new investment and long-term expansion was limited. Growth rates in some industries were in fact lower than in the pre-war period, and there is only limited evidence of new investment and diversification in manufacturing. In addition, despite its progress in producing consumer goods, the continent remained heavily dependent on Europe and North America for machinery and other capital equipment, and finishing materials such as chemicals and dyestuffs.

Whatever the uncertain impact of World War I, there is little doubt that the Great Depression did have a very positive role in stimulating industrialization. The period clearly revealed the vulnerability of export-dependent economies and the case for economic diversification and greater self-sufficiency, incorporating a more substantial manufacturing sector. This prompted moves towards import substitution industrialization (ISI), Latin America's first substantive move towards more autonomous industrial development.

By 1929, of the five most industrialized Latin American states, only Argentina derived more than 20 per cent of GNP from the industrial sector; in Mexico, Brazil, Chile and Colombia the contribution was between 5 and 15 per cent. The manufacturing capacity which existed had been largely prompted by the export sector, with some limited, generally small-scale development of industries meeting basic needs. The decline of the export sector and a matching fall in the capacity to import consequent upon the world slump prompted greater necessity for domestic industrialization. The process of ISI therefore began in the 1930s, was heightened during World War II and continued in the 1950s, particularly after its precise articulation by Raul Prebisch and the United Nations Economic Commission for Latin America. Rates of industrial production increased rapidly. Over the period 1929 to 1937 the increase was over 90 per cent in Colombia, over 40 per cent in Mexico and Brazil, and 23 per cent in Argentina; between 1937 and 1947 the growth rate was in excess of 70 per cent in all four countries.

These trends were also closely linked to emerging economic nationalism, in which the republics sought greater control over the nature and direction of their economic development, and greater independence from the perceived influences of Britain and the United States. This was also a period in which nationalist and populist governments were influential – Getulío Vargas in Brazil (1930–45), Lázaro Cardenas in Mexico (1934–40) and, slightly later, Juan Perón in Argentina (1946–55). The case for developing manufacturing obtained greater acceptance, and was fostered by the imposition of tariff barriers, import controls, subsidies and infrastructural provision.

Such trends saw increasing direct or indirect involvement by the State in the economy and in the manufacturing sector. Governments began to provide tariff protection for consumer goods industries such as textiles, clothing, food and household goods. Conversely, to encourage industrial expansion, concessions were granted on the import of machinery and essential raw materials. Government loans became available for industry, and industrial banks to provide financial support were established. Towards the end of the period tentative steps towards formal economic planning were begun, but even in the 1930s frameworks for development were created. In Chile, following the establishment of mining and industrial banks in 1927 and 1928 respectively, a development corporation, the Corporación de Fomento de la Producción was established in 1939. This used domestic funds and loans from the Export-Import Bank of Washington to formulate an electrification plan and encourage the development of the steel, wood and paper, sugar, ʰ ˀhemical industries. Similarly in Mexico the Nacional Finan   ʌted in 1934, provided funds for infrastructural and indus.   .s, including the Monclava steelworks. This latter interest reflected, as in the case of oil nationalization, a growing view in Latin America that certain sectors of the economy, including mining, fuel and power production, steelmaking and some strategic industries should be subject to some degree of government control or participation.

However, not all of the new initiatives came from government. Foreign companies continued to maintain established interests in plantation agriculture, energy and transport, and mining, but also began to invest in directly productive manufacturing activites. The latter frequently formed part of an evolutionary process, from the import of finished goods into Latin America, through bulk import and local packaging, local assembly of imported components, to local manufacture using a combination of local and imported inputs, or entirely local raw materials and components. Although such foreign involvement was of relatively limited scale, compared to public and private domestic capital, it was frequently concentrated in certain sectors of significance for the development and diversification of the industrial sector, such as the vehicle assembly, electrical, chemical and pharmaceutical industries. Such developments were associated particularly with American and British capital, but also French, German, Belgian, Italian and other European countries. It involved firms which would now be termed 'multinationals' such as Imperial Chemical Industries, British-American Tobacco Co., Coats, Parke Davis, Lone Star Cement Co., Du Pont, Ford, General Electric, Bayer, Nestlé, and Pirelli.

It should not be overlooked, however, that besides government and foreign investment, a considerable impetus to industrialization came from private domestic capital. Estimates for the late 1930s suggest that in Chile, Colombia and Uruguay for example, the bulk of manufacturing plants were domestically owned. It was generally the case though that domestic control was of greater significance in the more traditional industries, and involved lower levels of capitalization and less mechanization.

By the late 1930s there were three broad elements in Latin America's industrialization. The first was the continuing existence of a sector concerned with the processing or semi-processing of commodities for export such as sugar, meat, metals and petroleum. Foreign ownership remained significant in at least some of these activities, which were often associated with capital intensive technology and located at the point of production (which might be an 'enclave' of development in an otherwise little developed area) or at the ports. Secondly there were the 'traditional' industries, stimulated by the ISI process and becoming larger in scale, and with a significant degree of domestic capital – in food, drink, textile, clothing and building material production. Thirdly there were 'new' industries beginning to develop, representing the diffusion of North Atlantic growth industries, producing consumer durable and intermediate goods – vehicles, electrical goods, machinery, chemicals, and the like. Two other elements might also be noted. There was the establishment of heavy industry in those countries with adequate resources, markets and levels of development, so that by 1944 steel production in Brazil, Mexico and Argentina was 221,000 tons, 175,000 tons and 150,000 tons respectively. At the other extreme there was still the survival of a small scale craft tradition, which was little mechanized and frequently seasonal, providing basic needs in the small towns and the countryside – grain milling, simple textiles, bricks and tiles, household furniture and other essential items.

This outline, however, represents the general pattern of Latin American industrialization which could be observed in the 1930s and 1940s. There were considerable internal variations in the level of industrialization, the structure of industry and its spatial patterns. As Fig. 4.3 indicates, in the early post-1945 period there were considerable contrasts in industrial development, as measured by manufacturing employment, with Brazil, Argentina and Mexico having the largest work force, with Colombia, Chile, Peru (381,000 workers in 1940) and Cuba forming a second group, and most of the other republics with less than 100,000 workers. The map also indicates the contribution of industry to GDP, but is slightly misleading in that the data relate not only to manufacturing, but also mining, and gas, water and electricity production. The mining element accounts for the apparently high level of industrialization in Venezuela, Chile and Bolivia. If this limitation is taken note of, it is evident that the level of industrialization in 1950 was low throughout Central America and in Colombia, Ecuador and Paraguay. The 'more industrialized' countries were Argentina, Brazil, Mexico, Peru and Uruguay.

Figure 4.4 gives some indication of the variable pattern of industrialization where, by 1945, only Brazil, Mexico, Chile and Argentina were producing pig iron, while most of the republics had brewing industries. This pattern demonstrates the considerable internal variation within Latin America in levels of industrial development; it also indicates the varying nature of the development in that only a few countries had, by 1945, been able to develop more basic and modern industrial sectors: for most countries industrialization was largely restricted to export processing activities and the production of essential consumer goods. It can be noted, for example, that in addition to iron and steel, production of basic chemicals in 1945 was limited to such countries as Argentina, Chile, and Peru. A number of countries, including Brazil, Argentina, Mexico, Peru, Colombia, Ecuador, Venezuela, Chile and El Salvador had substantial cotton textile industries. The capacity of the Brazilian industry was greater than that of Canada, and those of Mexico, Argentina and Peru larger than that of Australia; however, only Brazil, Argentina and Colombia had begun to produce artificial and synthetic fibres. Processing or semi-processing of raw materials for export gave some restricted industrial activity, in the mineral processing and metalworking industry, with production of refined copper, lead and zinc in Mexico, copper and lead in Peru, copper in Chile and lead in Argentina.

There was thus a pattern of uneven levels of industrial development by the 1939–45 period, and considerable contrast not

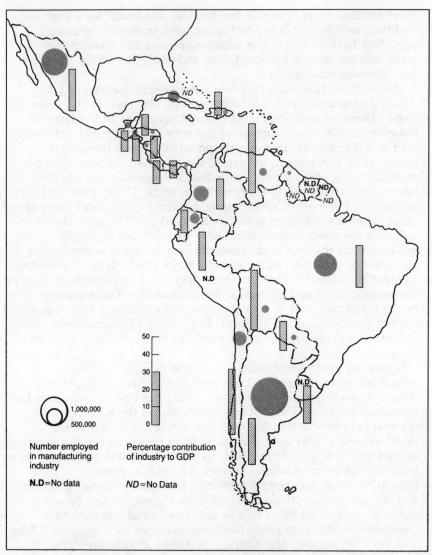

Fig. 4.3 The contribution of industry to employment and GDP *c.* 1950
(*Source*: From Mitchell B R (1983) *International Historical Statistics:
the Americas and Australasia*. Macmillan, London)

only in the scale but in the structure of development. Most countries
were restricted to a small scale, narrowly based manufacturing
sector; only a small number had a more substantial and diversified
industrial structure, and even this, in general, tended to be of
limited significance on the world scene. Furthermore, there were
also considerable contrasts within individual countries in the scale
and structure of their industries. Indeed the national picture tended

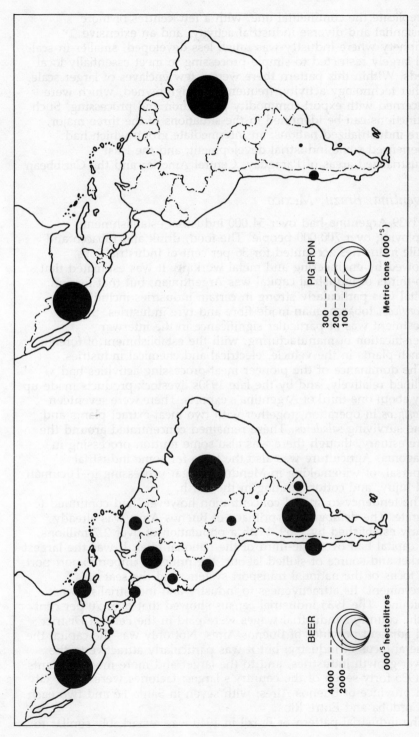

Fig. 4.4 Distribution of the production of two industrial commodities, 1945 (*Source:* From Mitchell B R (1983) *International Historical Statistics: the Americas and Australasia.* Macmillan, London)

to replicate the continental one, with a few centres of more substantial and diverse industrial activity, and an extensive periphery where industry was much less developed, smaller in scale, and largely restricted to simple processing to meet essentially local needs. Within this pattern there were a few enclaves of larger scale, higher technology activity, frequently foreign-owned, which were concerned with export-commodity production and processing. Such distinctions can be identified in the situations of the three major, more industrialized nations; an intermediate group which had experienced some industrial development; and the least industrialized areas of Paraguay, Central America and the Caribbean.

Argentina, Brazil, Mexico

By 1939 Argentina had over 54,000 industrial establishments, employing over 700,000 people. The food, drink and tobacco and textile industries accounted for 36 per cent of industrial jobs, followed by engineering and metal working. It was estimated that two-thirds of industrial capital was Argentinian, but that foreign capital was particularly strong in certain industries including the *frigoríficos*, tobacco, man-made fibre and tyre industries. Such investment was of particular significance in the interwar diversification of manufacturing, with the establishment of foreign branch-plants in the vehicle, electrical and chemical industries.

The dominance of the pioneer meat-processing activities had declined relatively, and by the late 1930s livestock products made up only about one-third of Argentina's exports. There were seventeen *frigoríficos* in operation, together with two meat-extract plants and some surviving *saladeros*. These remained concentrated around the Plate estuary, though there was also some mutton processing in Patagonia. Agriculture was also the basis for some industrial dispersal, of winemaking in Mendoza, sugar processing in Tucumán and Jujuy, and cotton ginning in the north.

The tendency to spatial concentration however, had continued to operate. The demographic primacy of Buenos Aires was already clearly established by 1939. With a population of over 2.7 millions, the capital had over one-third of the national total. It was the largest market and source of skilled labour, the principal import-export port, the focus of the national transport system, and the seat of government. Its attractiveness to industry and industrialists was profound. The 1935 industrial census showed that over 70 per cent of the country's industrial wages were paid in the Federal District and adjacent province of Buenos Aires. Not only was the capital the general focus of industry, but it was particularly attractive to the newer growth industries, and to the larger and more modern plants. In 1935 forty-seven of the country's largest factories were in the city and province of Buenos Aires, with seven in Santa Fé and two each in Cordoba and Entre Rios.

The industrial pattern of Brazil in 1940 was remarkably similar to that of Argentina. There were 41,000 manufacturing plants and 815,000 industrial employees, with the textile and clothing and food, drink and tobacco industries providing almost 60 per cent of the jobs. The next most important employers were the metallurgical, non-metallic mineral, and chemical and pharmaceutical industries.

Spatial concentration was also marked and increasing. In 1920, 63 per cent of Brazil's industrial jobs were in the city and state of Rio de Janeiro, and in Minas Gerais and São Paulo states; by 1940 this proportion had risen to 66 per cent. In this case however the capital city was not pre-eminent. Instead the city and state of São Paulo, the focus of the coffee boom, had come to dominate the country's industrial geography. In 1940 over one-third of industrial jobs were to be found there, and the state was the principal concentration of the engineering, vehicle and electrical industries. Foreign capital was significant in such sectors and in 1940 accounted for 27 per cent of the nominal manufacturing investment. The State had, however, become rather more involved in the economy than was the case in Argentina. It had given support to some areas of primary production and to the exercise of control over public utilities. During World War II it also became directly involved in the mining of iron ore and the production of iron and steel, lorries and chemicals.

Away from the emerging industrial 'core' of the south-east, development was much more limited. The north-east region in 1940 had some 145,000 workers; but over 70 per cent were employed in the traditional textile, clothing, food, drink and tobacco industries.

Similar broad patterns characterize the Mexican industrial economy of the late interwar period, though with some modifications introduced by a greater role for the State, consequent upon the revolution of 1910–17, and of foreign capital, deriving from proximity to the USA. Early objectives of the Mexican Revolution had been concerned with agrarian reform and various social and political programmes, with considerable involvement by the federal government in the economy. In the late 1930s and early 1940s more attention was given to the manufacturing sector, and financial and infrastructural development provided a basis for development. By 1940 there were almost 333,000 workers in mining and manufacturing. In Mexico's case the mining sector was of considerable significance, both directly and as basis for industrialization, with over 16 per cent of the work force in the metal mining and oil industries. In manufacturing *per se* textiles and food were the principal industries.

By 1939 Mexico City and the Valley of Mexico were well established as the principal focus of the country's industry. However, although that pre-eminence is clear, there is an element of dispersal of industrial activity, linked primarily to Mexico's mineral resources. The petroleum resources of the Gulf Coast provided the basis for oil refining and processing activities, with three refineries around Tampico and one at Minatitlan. Similarly the mining of lead, copper, zinc and other minerals provided the basis for smelting and semi-processing activities in San Luis Potosí, Zacatecas, Coahuila and other northern states.

The intermediate group

The experience of what might be termed the middle group of Latin American countries tended to be smaller in scale and later in the development of an industrial sector, but with some features in common with the larger states. Chile provides a nice example of spatial and structural contrasts in its industrial pattern. In 1940 its principal manufacturing industries (excluding the handicraft sector) employed some 116,000 people, with food and textiles the largest sectors. The decline of the nitrate industry, the main prop to the economy until the 1920s had led to considerable interest in industrialization, as an alternative economic base, encouraged after 1939 by the creation of the Corporación de Fomento. This had given the State a considerable direct and indirect role in manufacturing, fostering development in the chemical, pharmaceutical, steel, metalworking and tyre industries. An important element in the role of the Corporación was its involvement in the steel industry. It provided half the capital for the Cia de Acero del Pacífico, announced in 1943 and opened in 1950 at Hauchipato near Concepción. This has provided a point of industrial growth in the centre-south, with several subsequent expansions of the plant, and the establishment of local steel-using industries. The plant was mainly dependent on Chilean raw materials, and intended to make Chile self-sufficient in iron and steel. Its establishment also reflected a common view of the period that iron and steel capacity was essential to the industrialization process.

The principal concentration of industry in Chile, however, emerged around the capital Santiago and its port Valparaiso, the two roles being major attractions for ISI consumer industries serving the domestic market. Thus, by 1945 the bulk of the food, drink and tobacco, clothing, furniture and similar industries was located around the Santiago–Valparaiso axis.

The third element in the country's industrial geography in the 1930s was a characteristically Latin American one of an export-orientated processing activity, in the north. Metallic mining, for copper, gold and silver, had been carried out in this region before the nitrate boom, but only with the decline of the latter did copper return to significance. This was consequent upon technological advances in the developed world, which made possible the exploitation of low grade ores. This exploitation began in the second decade of this century, with the mines of Chuquicamata, El Teniente and Potrerillos being developed by the American Kennecot Copper Corp. and Anaconda Co. This involved not only massive mining operations but also processing plants, adjacent to the mines or in the exporting ports. This has generated nodes of development and settlement in an otherwise harsh environment, giving nucleations of mine, processing plants, service activities and townships sustained by external linkages and demand.

Variations on this industrial structure pattern emerged in most of the other middle group countries. In Peru the capital/port of Lima-Callao provided the focus of the bulk of the consumer industries. Only the pull of raw materials offered a modification of the patterns with sugar *centrales* located close to the plantations of the coastal valleys. Cotton ginning and some small mills were similarly located close to the cotton fields, but the major plants were close to Lima and newer industrial centres had emerged at Arequipa and Trujillo. The other principal industrial focus was associated with mining and metal refining for export. Cerro de Pasco, over 4,000m (13,123 ft) up in the Andes, had provided silver for the Spaniards and during the nineteenth century. The construction of the Lima-Oroya-Cerro de Pasco railway (1870–1904) permitted the exploitation of copper by the American Cerro de Pasco Corp. The nature of this type of large scale foreign mining activity is indicated by the corporation's operation, in the early 1940s, of five mining camps, three concentrating mills, a major smelter, a coal mine, coke ovens, four power stations, brickworks, three chemical plants and 176 km (110 miles) of railway!

Colombia and Venezuela had made less industrial progress by the early 1940s, constrained by market size, low income, difficult terrain and limited transport. Despite the early stimulus of the coffee boom, in 1920 there were just over a hundred factories in Colombia. A doubling of coffee exports between 1920 and 1929 provided new investment capital and the economic crisis of 1931 prompted ISI through government protective measures, such that by 1942 over 1,400 industrial establishments were recorded. Many of these were small however, with only 158 employing more than fifty workers, and 878 employing less than ten. Pre-1945 industrialization in both countries was largely limited to the consumer sector. The substantial expansion of Venezuelan oil production from 1928 provided a potential basis for prosperity and industrial investment, but this was not utilized until after 1945, so that the country continued to depend on imported manufactures to a large extent. Oil did provide the basis for some pre-1945 refining capacity on the Lake Maracaibo and eastern fields, but the bulk of the oil was exported as crude. During the Second World War exploitation of iron ore by the Bethlehem Steel Corp. began, to provide a later growth point in Venezuelan Guyana. However, aside from these mineral-based activities, the focus of Venezuela's limited industry was Caracas. Colombia's pattern was somewhat different, and indeed untypical of Latin America, with a greater degree of dispersion, and lesser significance for the capital city. In 1945 Medellín was the leading industrial centre (22 per cent of jobs) followed by Bogotá (17), Barranquilla (10) and Cali (7).

Industrial activity in Ecuador and Bolivia was limited in scale and variety. In the former, over a third of an estimated 10,000 industrial workers were employed in the textile industry, and food, drink and clothing were also important employers. Refining of petroleum at the Santa Elena oilfields provided an element of diversity and dispersion, but the principal industrial centres were the capital, Quito, and the main port Guayaquil. Craft industries such as wool textiles and leather goods provided a limited element of dispersion, as did the production of 'Panama' hats from palm leaves near the coastal town of Jipajapa and Monticristi and in the highlands where cheap labour was more abundant. Limited development, lack of diversity and metropolitan dominance also characterized Bolivia, with food, drink, textiles and clothing providing much of the employment, and La Paz generating over three-quarters of industrial output. Limited markets, acute poverty and difficult terrain have been major checks to industrial development. The mining of tin, from the last decade of the nineteenth century, provided a crucial element in the Bolivian economy and an element of diversity. Tin exports provided over half of government revenue. In 1939 production from the Potosí, Oruro and Uncía areas was largely controlled by three large corporations, but the tin ore was not smelted within the country and there was therefore little industrial spin-off from this major economic activity.

The least industrialized countries

The remaining countries to be considered have in common small area (all, except Paraguay, being below 130,000 sq km (50,000 square miles)), small population (all, except Cuba, having less than 3 million inhabitants circa 1950), and a marked dependence in the nineteenth and early twentieth centuries on one or two export commodities. The latter were primarily agricultural – coffee, bananas, sugar, tobacco, livestock products and similar commodities. In some of these activities control was exercised by a few, foreign companies. Moreover, in some cases the product generated little or no local processing activity.

Small size frequently implied a restricted resource base, and the low population and low income were major constraints on any industrial development. In the case of the six Central American Republics, which in 1950 had a combined population of little more than 8 million people, coffee and bananas became dominant elements in their economies. Coffee cultivation developed from the 1850s to become of major importance in Costa Rica, El Salvador, Nicaragua and Guatemala. Banana production for export began in the 1880s on the humid, fertile Caribbean lowlands, but became of major significance as a large scale plantation crop with the creation of the United Fruit Co. in 1899 and the Standard Fruit and Steamship Co. in 1924. These companies were responsible for the massive transformation of the lowlands, creating plantations, drainage canals, railway nets, company towns and banana ports. Banana production from the area reached an interwar peak in 1930 when 48 million stems of bananas were shipped. These developments created elements of progress and prosperity, but also acute dependency on a single crop – and company – as 'banana republics'. This dependence was both vulnerable and ephemeral, since disease led to many of the Caribbean plantations, and their associated paraphernalia, being abandoned by 1945, with exports falling below 20 million stems in the early 1940s. Coffee and bananas generated little or no industrial activity, so that prior to 1945 such industry as there was, was confined to cottage industries and a limited amount of food processing – coffee, sugar, flour, alcohol, and clothing and textiles. In the early 1940s the six countries had a total of fourteen cotton mills. Such activities were concentrated in the principal cities – Guatamala City, San Salvador, Tegucigalpa, Managua and San José.

Farming also provided the basis of Uruguay's economy and exports prior to 1945, with pastoral products responsible for 80 per cent of exports. The dominant elements were wool (40 per cent) and frozen, chilled and canned beef (20 per cent). The latter contributed an important industrial element, with four *frigoríficos* (three foreign and one state-owned), and the Liebig meat extract plant at Fray Bentos. Pastoral raw materials also provided the basis for woollen mills and leather goods. These, together with other food processing and other light industries, provided the basic industrial structure. In the interwar period Uruguay did generate a significant industrial sector – the 1936 industrial census recorded about as many people working in industry as in agriculture. There were over 105,000 people in manufacturing, though mostly in small units of production; of over 11,000 establishments, only ninety-nine employed more than a hundred people. The markedly primate Montevideo (770,000 inhabitants of a national total of 1.97 million) dominated the location of industry, except for the meat-processing plants.

Paraguay, by contrast, was the poorest, least populous and least industrialized part of South America in 1945. Limited resources and debilitating wars in 1860–75 and 1932–35 provided little basis for economic advance. There was a heavy dependence on subsistence agriculture, together with some cash production of cotton, sugar, bitter oranges and livestock. Yerba maté and quebracho extract were also export products. Manufacturing was largely confined to processing some of these products – cotton gins, two cotton mills, eleven sugar mills, a flour mill, four quebracho extract plants and three meat product plants!

In the Greater Antilles the Cuban economy before 1945 was dominated by sugar and to a lesser extent tobacco. The former was subject largely to American ownership and quotas in the US market. Peak interwar production was 5.2 million tons in 1928 but this was almost halved by 1939, and the processing industry stagnated, with no new capacity or modernization after 1927. Tobacco production also declined, with a consequent fall in the number of essentially small-scale tobacco-using factories. There were some other food processing activities, together with textiles, building materials, furniture and leather goods. The Dominican Republic also depended on sugar, produced on a large scale, and on cacao and coffee, produced by small farmers. In 1940 sugar accounted for almost half of industrial output, and other food products for a further 25 per cent. The remaining sector was limited to textile goods, tobacco, leather and similar products for local use.

Though the industrialization process cannot be interpreted in stage terms, it is evident that there was a considerable range in the level and nature of industrialization in Latin America by 1945. Some countries had developed a fairly substantial and diverse industrial structure with a range of consumer durable and non-durable goods, and some elements of basic industry. Others had made less progress, and their limited industrial capacity was largely confined to the essential production of food, clothing and building materials. A significant element throughout most of the continent was the legacy of the colonial and neo-colonial experience, of an export-processing activity, based on agricultural and mineral products aimed at developed world markets, and frequently controlled by foreign capital, and utilizing advanced technology.

The potential role of manufacturing in economic development and diversification had been recognized, even if progress towards industrialization was variable and generally still rather limited. Most countries had scarcely begun the process of import substitution industrialization, though there was an emerging investment mix of domestic private and State capital and an important foreign element in the 'traditional' export and 'new' growth industries. Spatial concentration of industrial activity was already noticeable in most countries, with important implications for later urbanization processes and regional development patterns. Away from these cores the export processing activities provided 'enclaves' in extensive rural or undeveloped peripheries from which industry was entirely absent or confined to traditional crafts and simple processing of foodstuffs and other local necessities.

It must also be noted that many industries remained heavily dependent on imported equipment and machinery. From the earliest attempts to create 'modern' factories in the late nineteenth century there had been a necessity to import the technology of the already industrialized countries. Although some local production of equipment had begun in the interwar period, frequently by branch plants of European and North American firms, much was still imported. In consequence of technical progress in the industrialized world, such machinery was increasingly capital intensive, requiring less labour to produce a larger supply of goods; yet for Latin America in this phase of industrialization, capital was relatively scarce and labour relatively cheap and abundant. Thus, whatever the contribution of manufacturing to economic growth and diversification, its perceived role in the absorption of surplus labour was restricted, a paradox of increasing significance in the period after 1945.

Further reading

Bernstein M D (1964) *The Mexican Mining Industry 1890–1950*. State University of New York, Albany. (A detailed study of a crucial sector in the Mexican economy for the pre- and post-revolutionary periods.)

Crossley J C (1976) 'The location of beef processing', *Annals Association of American Geographers* **66**, 60–75. (A theoretical and empirical study of the evolution of the meat trade between Europe and the Pampas.)

Finch M H J (1981) *A Political Economy of Uruguay since 1870*. Macmillan, London. (A thorough survey of the evolution of the Uruguayan economy and the internal and external factors involved.)

Furtado C (1970) *Economic Development of Latin America: a Survey from Colonial Times to the Cuban Revolution*. CUP London. (A very useful and thorough overview of the development process in Latin America. Probably the best general introduction to the topic.)

Leff N H (1982) *Underdevelopment and Development in Brazil*, (2 vols). Allen & Unwin, London. (A detailed examination of development process in Brazil 1822–1947, with some exploration of the emergence of regional imbalances. Explores a range of potential influences on both process and pattern and provokes ideas for further research.)

Morris A (1981) *Latin America: Economic Development and Regional Differentiation*. Hutchinson, London. (A useful introduction to the development of Latin America at a broad scale.)

Platt D C M (1972) *Latin America and British Trade, 1806–1914*. Black, London. (Detailed economic history of the changing nature of the nineteenth-century trade of Latin America, in which Britain played a major role.)

Thorp R and **Bertram G** (1978) *Peru, 1890–1977: Growth and Policy in an Open Economy*. Macmillan, London. (A comprehensive survey of economic development and change.)

Versiani F R (1979) 'Industrial investment in an 'export' economy: the Brazilian experience before 1914', *University of London Institute of Latin American Studies Working Paper*, No. 2. (A brief but detailed monograph of the emergence of Brazil's cotton manufacture and trade.)

CHAPTER 5

Modern manufacturing growth in Latin America

Robert N. Gwynne

With the exception of Cuba and Nicaragua, the economies of Latin America are closely interlinked into the world capitalist system. The operation of the world capitalist system significantly affects both the pace and rhythm of economic growth in each Latin American country. Recession in the developed world has major repercussions on the Latin American economies as the 1979–83 recession demonstrated. Economic growth in the developed world can have similarly advantageous results, providing export-led growth in Latin American economies. Generally speaking, the more 'open' an economy is to outside influences from the world economy, the more it will benefit in times of world economic growth and the more it will suffer in times of world recession. An excellent example of this can be seen in the recent economic history of Chile. Between 1974 and 1977, Chile became one of the most 'open' economies in Latin America and between 1977 and 1981 the economy grew at an average of 8 per cent a year. With the world recession hitting Chile in 1981, there followed two years of widespread economic decline – a decline, it should be added, exacerbated by poorly judged exonomic policies. The economy of each Latin American country is then inextricably linked to the performance of the world economy as a whole and in particular to that of the developed world.

It is best to see modern manufacturing in Latin America as the result of both exogenous and endogenous forces. This is because modern manufacturing in Latin America has been fashioned both by mechanisms that originate outside the country in question and by economic processes very much internal to that country. For this reason, this chapter will first look at the external mechanisms that affect Latin American industry before examining some of the major sectoral and spatial characteristics of modern manufacturing in Latin America.

External mechanisms

Background

As John Dickenson has shown in the previous chapter the origins of
Latin American industrialization go back to the late nineteenth
century. However, government influence on industrialization was
generally small. This changed with the onset of the world
depression after 1929 as governments increasingly became involved
in the management of the economy as a whole and industry in
particular. The initial involvement of government was the result of
balance of payments crises caused by the precipitate decline of Latin
American exports from an average of about $5,000 million in 1928–29
to $1,500 million in 1933. With the concomitant decline in foreign
exchange, Latin American governments found themselves able to
finance decreasing amounts of imported manufactured goods from
the industrial countries. As a result various measures were taken to
conserve and ration decreased foreign exchange resources. Tariffs
were raised by most countries, import quotas were enforced and
restrictions placed on the use of foreign exchange. An array of
policies was developed out of the crisis, and an administrative
machinery established to carry them out. In general terms, Latin
America changed from a set of free-trade economies to one of highly
protected economies. Tariffs, quotas, and exchange controls provided
protection from foreign competitors by making the entry of foreign
goods expensive or impossible. Latin American entrepreneurs,
observing the scarcity of goods and the level of protection, began to
produce or increase the production of goods previously imported.
Industrial production and employment increased as a result. In
Chile, industrial employment increased by over 5 per cent a year
between 1928 and 1937, but in those sectors which protection
particularly favoured, employment rose at much higher rates – 10
per cent per annum in metal products, 16 per cent in textiles and 19
per cent in chemicals.

Such a strategy of industrial development behind high protective
tariffs continued to be followed in most Latin American countries
after the adverse effects of the Depression had diminished. It was
noted that all major industrial countries had industrialized behind
high protective tariffs. It was only after a country had developed a
mature industrial structure that it could become involved in the free
trading of industrial goods. Protective policies also promoted the
development of infant industries, so that national firms could learn
new technologies and new processes without too much fear of
making large losses. Protection towards industry was also favoured
for reasons of employment. Industrial goods that are produced
nationally employ labour whereas imported goods obviously do not.
Furthermore, employment in manufacturing has both higher
productivity and wages than in other sectors of the economy so that

protective policies could radically change the labour market and increase the national product.

Due to such perceived advantages of protection for industrial development, protective policies were maintained after the Second World War and in the 1950s became formalized in the policy known as 'import substitution'. Adherents of the policy came to envisage four stages of industrial production within this process of industrialization. The first stage saw the production of basic non-durable consumer goods such as textiles, foodstuffs and pharmaceuticals. This was followed by the production of consumer durable products, such as cookers, radios, and televisions, and the critical motor vehicle industry: assembly began with a considerable ratio of imported parts. The third stage was critical in the industrializing process as it had to promote 'intermediate' industries producing the inputs for companies set up during the first and second stages. Typical industries at this stage would be chemical plants making paint, synthetic fibres, dyes and acids, or engineering works producing small motors and gearboxes, or parts industries for durable goods assembly. The final stage of the process would promote the development of the capital goods industry which would manufacture machinery and plant installations. It was the task of the government to plan and synchronize each subsequent stage in the process.

The promotion of manufacturing, then, can be seen as a government response to the vagaries of world trade and the need for what many governments perceived as a less dependent form of economic development. However, it could be argued that the resultant form of manufacturing growth through import substitution has left the sector even more dependent on the world capitalist system and the more developed countries in particular. Until recently there has been little internal development of technology and therefore a reliance on technology imported from developed countries. Furthermore, it has been argued that the import of such technology has not been in tune with local factor markets of labour and capital. In Latin America, the labour is abundant and capital scarce, but the transfer of technology from developed countries has been invariably capital-intensive rather than labour-intensive in nature. The organization of manufacturing, especially in high-technology sectors, is increasingly in the hands of foreign corporations. Meanwhile, national manufacturing companies have come to rely more on foreign capital in the 1970s. International banks were able to offer Latin American companies loans at low interest rates during that period, following the recycling of large OPEC surpluses after the oil price rises of 1973 and 1974. As a result the amounts that companies borrowed substantially increased. However, in the early 1980s, high interest rates and bankers' caution in lending further to Latin America have made investment capital once again scarce. Some companies who borrowed extensively in the 1970s have had to close because they have been unable to pay the increasing debts with which they are faced. In Chile, the two largest borrowers in the 1970s, the Cruzat-Larraín and BHC Groups, were faced with dollar debts greater than dollar assets after the 1982 devaluation of the dollar. In order to avoid widespread closures of productive plants and service agencies in the two largest Chilean conglomerates, the government had to intervene to run the two conglomerates. Smaller less strategic companies who borrowed too much in relation to their assets from the international banking system were simply declared bankrupt and terminated production.

International trade

Perhaps the most significant external mechanism that affects Latin American manufacturing is that of international trade. As the import substitution process progressed in Latin American countries, and production of consumer non-durables was followed by consumer durables, the less dependent form of economic development did not materialize. Indeed external constraints on industrial expansion often increased as balance of payments problems became more acute. This was mainly because import-substitution industrialization (ISI) only caused the *type* of import to change – from the finished product to the machinery and parts that made the product. Whereas when there was a balance of payments crisis in the period before ISI, it was feasible to curtail imports of the finished product, afterwards it was not. Machinery and parts needed to be imported constantly to keep factories producing and workers employed. ISI thus had the reverse effect of that anticipated. It increased the dependence of individual countries on the world trading system and decreased their room for manoeuvre in balance of payments crises and recessions.

In order to have more room for manoeuvre and to reduce the external constraints of their development some Latin American countries adopted a more outward-looking approach to industry in the 1970s. The exploitation of comparative advantage and the promotion of exports were dominant characteristics of this new approach. Small countries wished to overcome the problems of the small size of their market for manufactured goods. Larger countries wished to develop successful export sectors from industries previously only serving the domestic market.

Brazil has been very much in the forefront of reorientating the focus of industrial policy. The reorientation has been gradual, accelerating rapidly in the acute domestic recession of 1980 to 1982. The Brazilian policy has been one of maintaining as protected a market as possible whilst stimulating industrial exports by giving manufacturers an effective subsidy through a tax rebate. In 1981 this tax rebate (*Imposto sobre Produtos Industrializados*) was equivalent to a 15 per cent subsidy on export sales, although it was reduced to 9 per cent by 1983. As a result, the export of industrial goods (both manufactured and semi-manufactured) which constituted only about 10 per cent of total exports in 1971, now accounts for as much as 50

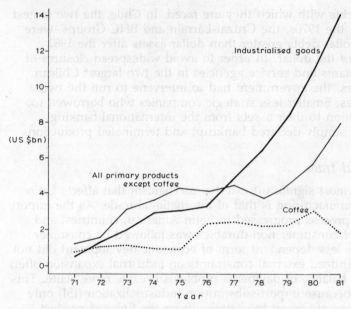

Fig. 5.1 Brazilian exports, 1971–81 (*Source*: Banco do Brasil)

per cent (see Fig. 5.1). In the 1980–81 domestic recession, exports of chemicals increased in value by over 100 per cent, black and white TV sets by 70 per cent, shoes and cutlery by about 50 per cent and motor vehicles by around 40 per cent. Motor vehicle and components exports now account for almost 10 per cent of Brazil's total exports, a fact that has made it a leading vehicle exporter, more important than the United States.

The most radical reorientation of industrial policy, occurred in Chile after 1974 under the Pinochet government. Chile left the Andean Pact in 1976 in order to pursue unhindered economic policies based on principles of comparative advantage and free trade and with concomitant reductions in tariffs to 10 per cent. Major restructuring of the industrial economy ensued with the large metal goods and machinery sector recording an employment decline of 28 per cent between 1974 and 1978 but with the food processing and paper sectors recording sizeable production increases and a rise in employment of 6 per cent for the period. However, between 1980 and 1981 manufactured exports declined due to an overvalued peso. At the same time cheap imports caused grave difficulties for most of the import substituting industry that had survived the free-trade policies of the Pinochet government. The resulting trade deficit ($2,458 million in 1981), the need to repay large loans with high interest rates and spiralling unemployment caused the Pinochet government to revise substantially its policies in 1983 and 1984. Exchange controls, tariff increases, exchange taxes operated to reduce consumer goods imports in 1983 to about one quarter of their

1981 level although industrial exports remained stagnant. With the formal expulsion of the monetarist 'Chicago' boys from the Chilean treasury in April 1984, it would appear that export-led industrial growth has been deemed a failure and a return to a more protected market has been engineered. Between 1983 and 1984, industrial production increased by 12 per cent with eight import-substituting sectors recording growth rates in excess of 25 per cent – china, machinery, furniture, electrical products, tyres, textiles, clothing and glass.

Other Latin American countries attempted to reorientate their industrial policies towards exports. Colombia and Peru followed the Chilean example and reduced tariffs to the 30 per cent level. The oil-exporting countries of Mexico and Venezuela have attempted to promote manufactured exports by subsidies, but high protection rates, overvalued currencies (until 1982) and an expanding domestic market have meant that most industrialists have not shifted their gaze from the national horizon.

However, despite Brazil's dramatic success in boosting industrial exports in the 1970s and despite Chile's ability to diversify its exports into semi-manufactures in the 1976–80 period, there are several external constraints that can affect such industrial policies attempting to promote exports. One major problem is that of a protectionist reaction from the developed countries. In the 1960s and 1970s the developed countries became more open to imports of manufactured goods from less developed countries. For example, between 1971 and 1976, the EEC, Japan and United States implemented the Generalized System of Preferences for imports from developing countries. These have certainly been important in facilitating the export of manufactured goods from less developed countries. For example, the EEC's Generalized System of Preferences covers only between 3 per cent and 4 per cent of the total imports from Latin America but as much as 30 per cent of manufactured goods imported from that region. Nevertheless, it has been argued that the impact of the GSP has been limited. Instead of complete, permanent and generalized tariff reductions, the GSP allows for partial, temporary and discretionary reductions confined to non-sensitive products. The GSP also allows for import quotas which have often proved very restrictive. Certainly the 'partial, temporary and discretionary' nature of the GSP makes it possible for the developed countries to reduce easily whatever benefits the GSP has provided for the developing countries. Furthermore, there is evidence of distinctly more protectionist stances by the developed countries in the 1980s. The non-signing of the Multifibre Agreement by the EEC in 1981 is one example of this.

Latin American countries could be doubly penalized by a more protectionist stance from the developed countries as they have never developed a special relationship with one of the groupings of the developed world. For example, it has no special relationship with the EEC whose markets for manufactured goods are open, without tariff or quota restrictions, to exports from the fifty-two developing countries of Africa, the Caribbean and the Pacific Islands who were signatories at the Lomé Convention. Japan's system of regional preference favours the Far Eastern countries while the United States has not promoted such a system – at least not until the 1981 special arrangements for the Caribbean area. In this way, Latin American manufactured exports can be prevented from entering the markets of developed countries. One example of this was the decision of the West German government, under pressure from its car unions, to deny entry to finished Brazilian Volkswagen vehicles even though Brazilian-produced Volkswagens were in some cases cheaper than domestically-produced vehicles.

For this reason, the majority of Latin America's manufactured exports goes to other less developed countries. In 1975, for instance, 90 per cent of all Latin American manufactured exports stayed in the region. Subsequently, Brazil's major thrust in its policy of export promotion of manufactured goods has been directed to other Third World countries. Volkswagen cars have been exported in large numbers to such countries as Nigeria, Philippines and Chile. However, such a direction has had adverse repercussions in the 1979–83 recession, in which Third World countries suffered more than developed countries. As a result, they severely restricted their manufactured imports which in turn affected Brazil's manufactured exports. This was the major reason for Brazil's decline in manufactured exports between 1981 and 1982. The more volatile nature of Third World markets for manufactured imports has become another constraint on the export promotion policy of Brazil and other Latin American countries.

A fourth potential problem is that a significant proportion of the exports of manufactured products from Latin America is carried out by multinational companies. Early 1970s data demonstrate that 30 per cent of Mexican, Colombian and Argentinian manufactured exports were made through multinationals and as much as 43 per cent of Brazil's. Multinationals export from Latin American countries when there are cost advantages for them. These cost advantages most commonly arise from either labour cost differentials (particularly for South-North trade) or from a need to use as much of the capacity as possible of the manufacturing plant (often the incentive for increased South-South trade). Due in particular to the cheapness of Latin American labour, multinationals favour the location of assembly-type industries in Latin American countries – industries that require large amounts of labour. However, as assembly-type operations become more and more capital-intensive, as with the increasing use of robots in motor vehicle assembly, there will be little advantage of locating where labour is cheap. As the development of labour-saving technology proceeds apace in the North and undercuts cost of production in the South, there will be increasingly more powerful arguments for multinationals to bring 'home' the production from overseas – or at least to limit severely the imports from the countries of the South, including Latin America.

Major characteristics of recent manufacturing growth in Latin America

Differentials between large and small countries

As ISI progressed through the 1950s and 1960s, it became clear that the policy was more suitable for large than for small countries. Industries in the smaller countries of Latin America came to be characterized by high costs and consequently high prices because their markets were too small.

High costs and high prices made possible by high protective tariffs were due to a wide variety of factors, but one key structural problem was the lack of economies of scale in producing for small markets, particularly in the critical second and third stages of the import substitution process. As production increases in any operation, unit costs will normally decline as fixed costs (e.g. plant and technology) will be spread over more and more units. The decline in unit costs with increased production will slacken and even out at some stage (known as the minimum efficient scale) before perhaps rising as output increases still further. Figure 5.2 shows the

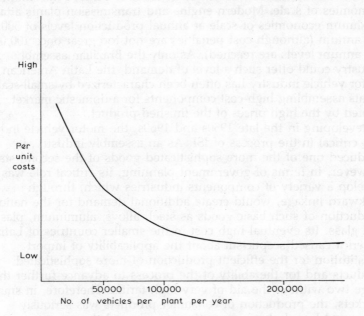

Fig. 5.2 Notional relationship between output of a vehicle assembly operation and per unit costs in Latin America *Source*: Gwynne (1985)

notional relationship between output of a vehicle assembly operation and per unit costs in Latin America. The minimum efficient scale is reached at approximately 200,000 vehicles per annum but there is a low per unit cost penalty for annual output between 100,000 and 200,000.

The example of the motor vehicle industry can demonstrate the relationship between economies of scale and actual production levels in Latin America. Table 5.1 shows vehicle output in each producing country by firm for 1977. It is evident that the assembly plant economies illustrated in Fig. 5.2 are only achieved by firms operating in Brazil (Volkswagen, General Motors, Ford and Fiat). Other countries had firms whose level of production was uneconomic. As a result, high-cost cars were being produced and sold to national consumers. A detailed cost analysis of the Chilean motor vehicle industry in 1969 demonstrated that cars were produced at up to three times the cost in the vehicle's country of origin. Cars in both Argentina and Venezuela were being sold at double international levels in the late 1970s.

Such high prices and high costs, were not solely the result of a lack of sufficient economies of scale in the assembly of vehicles. In order to promote certain stage three industries, many governments instituted schemes whereby assembly firms had to purchase increasing quantities of nationally produced components. However, components production can require an even greater level of production than vehicle assembly in order to achieve maximum economies of scale. Modern engine and transmission plants attain maximum economies of scale at annual production levels of 500,000 per annum (although cost penalties are not too great once 100,000 per annum levels are reached). As only the Brazilian assembly industry could offer such a level of demand, the Latin American motor vehicle industry has often been characterized by small-scale plants assembling high-cost components for a domestic market limited by the high prices of the finished product.

Developing in the late 1950s and 1960s, the motor vehicle industry was critical in the process of ISI. As an assembly industry, it produced one of the more sophisticated goods of the second stage. However, in terms of government planning, its critical role was to develop a variety of components industries which, through backward linkage, would create additional demand for the national production of such basic goods as steel, alloys, aluminium, plastics and glass. Its eventual high cost in the smaller countries of Latin America caused scepticism about the applicability of import substitution for the efficient production of more sophisticated products and for the ability of the process to advance further than stage two without the aid of very high tariffs. Therefore, in small markets, the production of consumer products was seriously hampered by a lack of economies of scale and the consequent high costs and low demand. Indeed, in times of stagnation, falling demand could start a vicious circle of lower economies of scale,

Table 5.1 Distribution of motor vehicle production by country and firm in Latin America, 1977

	Argentina	Brazil	Chile	Colombia	Mexico	Peru	Venezuela	Total	%
Chrysler	23,434	21,970	–	8,275	57,956	7,169	32,430	151,234	9.1
Fiat	47,837	77,963	3,120	4,128	–	–	4,889	137,937	8.3
Ford	56,795	130,197	–	–	50,503	–	61,665	299,160	17.9
General Motors	20,897	153,836	2,124	–	34,638	–	32,281	243,776	14.6
Renault	34,744	–	2,568	17,353	27,559	6,075	5,153	87,377	5.2
Volkswagen	–	472,192	–	–	52,143	–	5,340	535,750	32.1
Others	51,649	63,084	5,277	–	58,014	11,980	21,539	211,543	12.8
TOTAL	235,356	919,242	13,089	29,756	280,813	25,224	163,297	1,666,777	100.0
Percentage	14.1	55.2	0.8	1.8	16.8	1.5	9.8	100.0	

Source: Gwynne (1985)

higher costs and lower demand. With such high-cost production there were great difficulties in ever expanding the potential market through exporting.

For smaller countries, one possible solution to the restrictions of low demand and low economies of scale was for countries to group together and form, through a process of economic integration, a much larger market. The Central American Common Market (CACM) created in 1960 and consisting of Guatemala, El Salvador, Nicaragua, Costa Rica and Honduras gave manufacturers a market of 10 million, up to ten times larger than previous national markets. In 1969, five Andean countries (Chile, Bolivia, Peru, Ecuador and Colombia) formed the Andean Pact. The main motivator in its formation was Chile whose industrial development by import substitution had been severely constrained by small market size. The prospect of an enlarged market of 70 million attracted all five countries and later Venezuela in 1973.

What then has been the differential impact of import substitution policies on large and small countries and what have schemes of economic integration achieved for the smaller countries? The manufacturing growth rates presented in Table 5.2 basically demonstrate the influence of import substitution and economic integration policies on industrial development; the effects of the shift to more export-oriented policies of the late 1970s and 1980s are not evident. The large countries, where import substitution had the advantage of supplying large national markets, recorded much higher manufacturing growth rates than other countries, particularly in the 1965–73 period when average growth rates of 12 per cent in Brazil and 8 per cent in Mexico were achieved. The industrial achievements of the CACM countries from a low base are reflected in the 7.4 per cent average annual growth rate from 1965 to 1973, while the limited success of the Andean Group is demonstrated by the relatively low growth rate of 4.1 per cent between 1973 and 1978. Meanwhile, the failure of import substitution policies to provide industrial growth outside large countries or schemes of economic integration is most clearly seen in the River Plate countries where industrial decline was recorded between 1973 and 1978. Arguably, the import substitution process was 'completed' earlier in Argentina and Uruguay, but with the high costs of national industry preventing exports and further domestic expansion, such 'completion' brought industrial stagnation. Argentine industry declined by 1 per cent per annum between 1973 and 1978 causing per capita manufacturing GDP to fall from $531 in 1973 to $474 in 1978. In this way, the industrial weight of the River Plate region in Latin America fell from 35 per cent in 1950 to only 18 per cent in 1978 (see Table 5.3). Meanwhile, the two larger countries that had forged a more competitive industrial structure from import substitution, Brazil and Mexico, had increased their combined industrial weight in Latin America from 43 per cent to 62 per cent – a figure that effectively summarized the close links between market

Table 5.2 Latin America: Growth rates of the population and the manufacturing product, 1950–78

	Population		Manufacturing Product		
	1950–78	1950–65	1965–73	1973–78	1950–78
Large countries	3.1	7.3	10.4	6.3	8.0
Brazil	3.0	7.3	12.0	6.3	8.5
Mexico	3.3	7.2	8.1	6.3	7.3
Andean Pact (incl. Chile)	3.0	6.8	5.6	4.1	6.0
Bolivia	2.4	2.0	5.6	8.3	4.2
Chile	2.1	5.5	3.4	−1.4	3.7
Colombia	3.2	6.2	7.7	5.4	6.5
Ecuador	3.2	5.3	7.1	11.9	7.0
Peru	2.8	7.8	6.6	1.8	6.4
Venezuela	3.5	9.5	5.0	7.6	7.9
CACM* (incl. Panama)	3.0	7.4	7.4	5.3	7.0
Costa Rica	3.2	7.9	9.4	8.1	8.3
El Salvador	3.1	7.2	5.8	5.2	6.4
Guatemala	2.8	5.4	7.7	6.2	6.2
Honduras	3.2	8.3	6.4	6.3	7.4
Nicaragua	3.0	9.5	6.9	3.6	7.6
Panama	2.9	10.1	8.1	0.2	7.7
River Plate countries	1.6	4.6	5.6	−0.4	4.0
Argentina	1.6	4.8	5.9	−1.0	4.1
Paraguay	2.7	3.3	6.0	7.2	4.8
Uruguay	1.4	2.7	0.9	5.9	2.7
Latin America	2.8	6.3	8.2	4.5	6.5

Source: United Nations Economic Commission for Latin America
*Central American Common Market

size and industrial success in import substitution. Economic integration had little impact on the relationship. The industrial weighting of the twelve smaller countries involved in regional schemes of economic integration fell slightly from 22 per cent in 1950 to 20 per cent in 1978.

ISI has brought major structural changes in manufacturing. As would be expected from the import substitution model, industries producing non-durable consumer goods (e.g. clothing) have declined in relative importance as the industries producing intermediate goods, consumer durable (e.g. vehicles) and capital goods have developed. In 1950, non-durable consumer goods represented almost two-thirds of total manufacturing production, as opposed to 40 per cent today. In contrast, the relative importance of intermediate products in manufacturing output has risen from less than 25 per cent of the total to more than one-third at present. However, it is

Table 5.3 Latin America: Population, total and manufacturing gross domestic product, level of industrialization, industrial weight within the region and per capita manufacturing GDP, 1978

	Population thousands of inhabitants	GDP millions of 1970$	Manufacturing GDP millions of 1970$	Level of industrialization (manuf'g GDP as % GDP)	Industrial weight within the region (%)	Per capita manufacturing GDP (1970 prices)
Large countries	**184,898**	**170,140**	**48,172**	**28**	**62.4**	**261**
Brazil	119,477	101,056	30,327	30	39.3	254
Mexico	65,421	69,084	17,845	26	23.1	273
Andean Pact (incl. Chile)	**83,394**	**66,662**	**13,132**	**20**	**17.0**	**158**
Bolivia	5,848	2,072	325	16	0.4	56
Chile	10,843	10,335	2,451	24	3.2	226
Colombia	28,424	19,162	3,384	18	4.4	119
Ecuador	7,798	4,434	905	20	1.2	116
Peru	17,148	10,323	2,554	25	3.3	149
Venezuela	13,333	20,336	3,513	17	4.5	264
CACM (incl. Panama)	**21,002**	**12,279**	**2,198**	**18**	**2.8**	**105**
Costa Rica	2,111	2,031	461	23	0.6	218
El Salvador	4,524	2,238	436	19	0.6	96
Guatemala	6,623	3,783	617	16	0.8	93
Honduras	3,362	1,166	185	16	0.2	55
Nicaragua	2,559	1,195	238	20	0.3	93
Panama	1,823	1,866	261	14	0.3	143
River Plate countries	**32,490**	**43,042**	**13,761**	**32**	**17.8**	**424**
Argentina	26,395	38,011	12,512	33	16.2	474
Paraguay	2,888	1,553	250	16	0.3	87
Uruguay	3,207	3,478	999	29	1.3	312
Total Latin America	**321,784**	**292,123**	**77,263**	**26**	**100.0**	**240**

Source: United Nations Economic Commission for Latin America

the consumer durables and investment goods sector where the relative change has been most striking. This sector represented 11 per cent of total manufactures in 1950 and now accounts for more than one-quarter of manufactures produced.

Again, there are great differences among Latin American countries according to market size. Table 5.4 demonstrates that the process of import substitution has been much more complete in the large countries than in the smaller countries. Intermediate manufacturers and consumer durable/capital goods sectors accounted for 37 and 28 per cent respectively of total value-added in the larger countries (including Argentina) by 1975, against 35 and 17 per cent in medium-sized countries and only 26 and 9 per cent in small countries. The production of intermediate goods has expanded significantly in medium-sized countries but less so in small countries. Growth in basic metals is almost entirely restricted to the larger countries and the one medium-sized country, Venezuela, that benefits from considerable comparative advantages (in the form of cheap materials and power) in both steel and aluminium production at Ciudad Guayana. In 1976, 94 and 100 per cent of Latin American steel and aluminium production respectively corresponded to Argentina, Brazil, Mexico and Venezuela.

The most pronounced concentration of industrial growth in the larger countries occurred in the consumer durable/capital goods sectors. In terms of motor vehicle production, 86 per cent of assembly took place in Brazil, Mexico and Argentina in 1977. Meanwhile, the capital goods sector has 90 per cent of its production concentrated in the three largest countries, with Brazil having the technologically most advanced and broadest machinery sector.

Technology and size of country

The concentration of the production of intermediate, consumer durable and capital goods in Brazil, Mexico and Argentina is intimately linked not only with size of market but also with the internal generation and adaptation of new technology. Unlike the situation in the more developed countries, much of the process of modernization and technological change which takes place in Latin America consists of the imitation of products and processes that have already been developed in more advanced countries. At this stage of purchasing technological designs, there are few differences among Latin American countries. It is at the next stage, the adaptation and improvement of imported technology to meet local circumstances that the larger countries have considerable advantages over other countries. First, Argentina, Brazil and Mexico have a substantial number of highly-trained professional and technical personnel to generate national technological knowledge in addition to the technology that is imported. Second, the availability of such skilled personnel facilitates the adaptation of imported technology to local conditions. The Japanese mode of industrialization has

Table 5.4 Latin America: Structure of industrial production by size of country, 1950 and 1975 (Percentage of value added in the manufacturing sector)

		Non-durable consumer manufactures	Intermediate manufactures	Consumer durable and capital manufactures	Total manufactures
Large countries	1950	64	24	12	100
	1975	35	37	28	100
Medium-sized countries	1950	66	28	6	100
	1975	48	35	17	100
Small countries	1950	85	14	1	100
	1975	65	26	9	100

Source: United Nations Economic Commission for Latin America

demonstrated that modern industrial success is not so much rooted in the generation of totally new technologies as in the constant adaptation and improvement of imported technological designs – first to local conditions and later for export markets.

Although industrial technological expertise is concentrated in Argentina, Brazil and Mexico, these three countries have developed contrasting policies towards technology. Argentina has followed what could be termed a pre-Japanese model, stressing the autonomous development of technology and skilled personnel and reducing the impact of imported technology. Perhaps as a consequence, Argentina has been able to develop an autonomous technological sector closely linked to the government and military but unable to promote a dynamic private sector, partly because of restrictions placed on imported technology. The Mexican experience reflects a policy of concentrating technological investment in crucial areas or sectors in which Mexico has a comparative advantage (oil, petrochemicals, iron and steel, newsprint) and allowing the easy import of foreign technology in other sectors; this latter part is demonstrated in the crucial area of computers where no major Mexican firm has developed and where computers are either imported directly or simply assembled with minimal local content.

Meanwhile, Brazil has followed a policy similar to the Japanese model of developing technology. On the one hand, the import of new technological designs are welcomed but the domestic adaptation and improvement of these designs are vigorously promoted. The results of such a policy is reflected in the development of the national computer industry where a government body, the *Secretaria Especial de Informática* (SEI), has supervised its growth. The SEI has divided the industry into mainframe and smaller computers. Its policy towards mainframe computers has been to leave it to the multinationals, notably IBM and Burroughs, as long as they increase their manufacturing capacity in Brazil. The SEI's policy towards the manufacture of smaller computers is radically different. The government reserved this growth area almost exclusively for Brazilian manufacturers as far back as 1978. Four companies from the private sector were allowed to join COBRA – *Computadores e Sistemas Brasileiros* (a government-controlled company). The SEI allowed these companies to begin by buying technology abroad but approval of new projects depended on their use of Brazilian expertise and locally-produced components. In this way, the SEI stressed the need for these companies to adapt and improve technological designs, and to integrate their products into the Brazilian market and supply network. As a result, the five Brazilian companies have extended their range from minicomputers to microcomputers, and have been joined in this latter field by a number of new companies. By 1982, about twenty-five companies had approval for their products, and as many as fifteen had already begun to produce microcomputers, while many other companies have been created to produce printers, monitor screens, keyboards, disc-drives and other peripherals. One of the best-selling computers in 1982 was the Cobra 530, with 512 K bytes of memory, completely designed in Brazil and largely made from locally produced components.

Investment in research and development has a significant impact on the economic performance of the three countries. First, it has favourably affected manufacturing productivity. Second, research and development has begun to have a favourable impact on exports. The technological effort involved in adapting and improving upon an imported design in order to make it more suitable for local circumstances gives rise essentially to the emergence of a new product or process. The new product so developed may be more applicable for sale in third countries with similar local circumstances than the original imported design. Finally, with the development of new technologies and the adaptation of imported techniques being spatially concentrated in Brazil, Mexico and Argentina, it can be predicted that those three countries will play an important future role as suppliers of technologically sophisticated manufactures and technology itself through licensing arrangements and direct investment to neighbouring countries. In this way, the technological gap between the three larger countries and the rest of Latin America is likely to widen as the latest innovations and advances are rapidly transmitted to São Paulo and Mexico City where they are adapted for more general Latin American use.

The triple alliance

The process of industrial expansion in Latin America has been engineered through a distinctive institutional structure. This institutional structure has been referred to as the triple alliance – a triple alliance between state firms, national private enterprise and multinational corporations. The balance between these three institutional categories varies from country to country. Furthermore, the balance between these categories in any country is continually changing. In larger countries, national private enterprises have been losing some ground to both public enterprises and multinational corporations (in terms of their contribution to the industrial product). In smaller countries, on the other hand, the reverse is often true – national private enterprises are gaining ground at the expense of state firms and multinational corporations.

It has been the role of the state enterprise to invest in those intermediate and capital goods sectors that continued industrial expansion required. The state adopted this role because of inadequate domestic capital markets which meant that only the state could provide the necessary capital for such large investments. Furthermore, as some of the investment was of a strategic nature (power, communications, steel), governments wished to exclude the participation of the multinationals. State firms have also been established in the extractive industries and in the further processing and refining of the minerals concerned. State firms in this case have

Plate 7   A new town built to house workers at the Açominas steelworks in
Brazil. Industrial development at virgin sites necessitates the
provision of homes for migrant workers. (Photo: J. P. Dickenson)

developed due to a national wish for greater control over the crucial
resources of the country – resources crucial for exports, taxes, the
state budget, employment and the exchange rate.

The multinational enterprise in Latin America has historically been
of Western European and US origin but Japanese interests have been
increasing recently. The role of multinational enterprise as an agent
of Latin American industrialization is a crucial one, not least because
no other continent has received a major contribution from foreign
companies in their process of industrialization. The United States,
Japan and the countries of Europe have largely industrialized
through national agents, whether state or private. Other Third
World countries are presently industrializing with the assistance of
multinational enterprise, but Latin American countries have already
developed complex and diverse industrial structures from the
contributions of foreign investment.

Since the Second World War and the nationalization of
international mining interests in Mexico, Venezuela, Chile and Peru,
the flavour and structure of multinational enterprise has radically
changed. There has been a major increase in manufacturing as
opposed to mining investment by multinationals in Latin America
(see Table 5.5). Manufacturing investment has been particularly
significant in those countries that had reached the latter stages of ISI
– notably Mexico and Brazil. Thus 68.5 per cent of direct foreign

Table 5.5 Changes in the investment patterns of multinational enterprise in Brazil, Mexico and other Latin American countries

	1950				1978			
	Brazil	Mexico	Other Latin American countries	Total	Brazil	Mexico	Other Latin American countries	Total
Total investment (millions US $)	654	415	3,379	4,445	7,170	3,712	12,595	23,477
% of Total investment	14.7	9.3	76.0	100.0	30.5	15.8	53.7	100.0
Extractive (%)	1.1	29.9	30.1	25.8	3.7	2.6	10.3	7.1
Petroleum (%)	17.2	3.1	32.8	27.7	5.9	1.1	25.4	15.6
Manufacturing (%)	43.6	32.0	10.7	17.6	65.3	74.1	27.1	46.2
Public utilities (%)	21.2	25.8	20.2	20.9	0.4	0.6	2.1	1.3
Other (including finance and trade) (%)	16.9	9.2	6.2	8.0	24.7	21.6	35.1	29.8
Total	100.0	100.0	100.0	100.0	100.0	100.0	100.0	100.0

Source: Gereffi and Evans (1981)

investment in Latin American manufacturing was concentrated in Brazil and Mexico in 1978.

The majority of foreign investment is of US origin – 50 per cent in the large countries, 77 per cent in the medium-sized countries and 69 per cent in the smaller countries in 1976. However, both European and Japanese investments are increasing. Japanese investment is still most concentrated in the processing of raw materials (steel, aluminium) but European investment is shifting to high technology sectors such as motor vehicles and chemicals.

Indeed, there is increasing evidence to show that the rapid growth of multinational investment in manufacturing in Latin America (an increase of 146 per cent between 1967 and 1976) has taken place in the more dynamic and technologically innovative sectors. In the country with nearly 50 per cent of continental foreign direct investment in manufacturing (Brazil), 77 per cent of such investment was channelled into technologically dynamic sectors in 1976 – notably chemicals, vehicles, machine tools, pharmaceuticals, communications and the electrical and medical industries. This meant that in Brazil in 1979, multinational enterprise controlled 56 per cent of total assets in the transport sector, 51 per cent in the electrical and over 35 per cent in the machinery sector. These were the three sectors that recorded the highest growth rates during the 1970s in Brazil.

In contrast to the state firm and multinational enterprises, the national private firm is characterized by great diversity in terms of size, technological level and forms of organization. In most large and medium-sized countries, large national conglomerates have developed with a wide variety of manufacturing interests and often important tertiary functions in such areas as banking, insurance, finance, tourism, commerce and the media. At the other end of the size range, large numbers of small enterprises are at work filling the demand gaps left by or providing low-cost competition for the large state, national and international companies. Due to labour-intensive methods and low capital inputs, these enterprises generate a much higher proportion of employment than their production levels would indicate. In 1970, the 135,000 small firms of Brazil, while accounting for only 21 per cent of the total value of production, employed as much as 44 per cent of the total manufacturing employment. The organization of most of these small firms is simple, based around a single entrepreneur or family, and many are highly susceptible to changes in the macroeconomy or the policy shifts of the large companies. As a result, they have to be very adaptable and flexible in order to survive.

In terms of their relationship with multinational companies, it is often pointed out that the latter predominate in the technologically more dynamic sectors, leaving the national private firms to specialize in the more traditional industries. The 2,700 largest private firms in Brazil, for example, account for 75 per cent of production in the non-durable consumer goods sector but only 33 per cent and 45 per cent in the intermediate and metal goods machinery sectors.

However, such a distinction can be misleading. National private firms are active in some very dynamic sectors in Latin America. For example, in the motor vehicle industry, national private firms have specialized in components and parts production while leaving assembly and engine production to the multinationals. In Brazil, national firms figure prominently in the plastics, paper, machinery and microcomputer sectors. In Venezuela, the metal fabrication industry is dominated by such national firms as SIVENSA. Nevertheless, throughout Latin America, the majority of firms producing in the food, beverages, textiles, footwear, clothing, leather, cement, furniture and ceramic sectors are of national origin (see Table 5.6).

Table 5.6 Percentages of assets of largest 300 manufacturing firms in Brazil and Mexico held by US, other foreign and national companies, 1972

Industry	US companies (%)		Companies from other foreign countries (%)		National share (%)	
	Brazil	Mexico	Brazil	Mexico	Brazil	Mexico
Food	2	20	30	6	68	74
Textiles	6	0	38	5	56	95
Metal fabrication	4	48	21	8	75	44
Chemicals	34	54	35	14	31	32
Rubber	100	100	0	0	0	0
Non-electrical machinery	34	36	40	58	26	5
Electrical machinery	22	35	56	25	22	40
Transportation equipment	37	70	47	9	16	21
Total	16	36	34	16	50	48

Source: Gereffi and Evans (1981)

Latin American industrialization, in contrast to the industrialization of Europe, United States and Japan, has developed within the institutional framework of a triple alliance of state firms, multinational corporations and national private enterprise. Within the process of industrialization, each has developed a well-defined role. To a certain extent, the three contrasting roles have been compatible, particularly in the more mature industrial countries of Brazil and Mexico. Nevertheless, the fact that multinational enterprise is playing such a crucial role in the development of the more dynamic and technologically innovative sectors may have implications for the long-term stability of the industrial process – simply because Latin American governments are unable to influence

Plate 8 The Pemex refinery at Minantitlán. The oil fields of southern Mexico were mainly discovered in the 1950s and have provided the basis for associated refining and petrochemical industries.
(Photo: J. P. Dickenson)

the actions and decisions of multinational enterprise as closely as either state or private national firms.

Spatial concentration of manufacturing

The spatial concentration of manufacturing *between* countries in Latin America (i.e. in Brazil, Mexico and Argentina) has already been examined. Even more pronounced is the spatial concentration of manufacturing *within* countries – in the primate city, such as São Paulo, Mexico City, Buenos Aires and Santiago. The historical process of industrial concentration has already been referred to in John Dickenson's chapter. However, since the 1930s, the pattern of spatial concentration has become more and more pronounced. The concentration of Chilean manufacturing employment in Santiago, for example, increased from 43 to 57 per cent between 1928 and 1979. One important reason behind increasing concentration is that the new, growth industries are consistently drawn to the primate city. In 1967, the four growth sectors of pharmaceuticals, electrical durable goods, plastics and professional and scientific equipment had over 90 per cent of their Chilean employment generated in Santiago. With modern industries attracted to the primate city, the process of concentration continues to increase. In the two most advanced industrial nations of Brazil and Mexico well over 50 per cent of their respective manufacturing production and employment is located in the São Paulo and Mexico City agglomerations. Two-thirds of Argentina's industrial employment is located in the Buenos Aires metropolitan area, 75 per cent of Venezuelan value-added is generated from the Valencia-Caracas axial belt (see Fig. 5.3) and 70 per cent of Peruvian manufacturing employment is centred in Lima/Callao. In the smaller, less-industrialized countries where only the first stage of ISI has been completed, spatial concentration of industry in the primate city is greater still e.g. El Salvador.

How can such pronounced concentration of manufacturing employment and value-added be understood? As industrial entrepreneurs in the triple alliance compete for domestic markets and attempt to find export markets through offering cheaper products, locational costs and accessibility to markets are becoming dominant considerations. Such considerations normally reveal the advantages of the primate city.

As the preceding chapter demonstrated, the introduction of the railway in the nineteenth century created national transport networks that radiated out from the capital and/or primate city. As a result, Buenos Aires in Argentina, Montevideo in Uruguay, São Paulo and Rio in Brazil, Santiago-Valparaíso in Chile, Lima-Callao in Peru and Caracas-La Guaira in Venezuela adopted pivotal roles in the transport development of their countries. As a result, the most accessible point in each country was not the geographical centre but a point near or on the coast. As countries began to embark on policies of ISI, the large coastal city (or city at the centre of the nation's transport system) provided the most advantageous location for the development of three significant industrial types:

1. Consumer-goods industries that supply the whole of the national market from one or two plants.
2. Industries in which various raw materials from the hinterland of a country are processed and combined together in one place. This is because the primate city will be the only city linked to the various sources of the raw material.
3. Industries that have an element of imported raw materials and/or components in their productive structure.

These industrial types have been particularly prominent in the process of ISI. The original idea of ISI was to substitute imports of a product through the creation of a small number of national plants (Type 1). As the process of ISI continued, plants producing consumer durable products were established. These plants needed to import machinery and components, at least initially, for their manufacturing process (Type 3). Later the manufacture of intermediate products was encouraged, combining either raw materials from the country's hinterland or from abroad (Types 2 and 3).

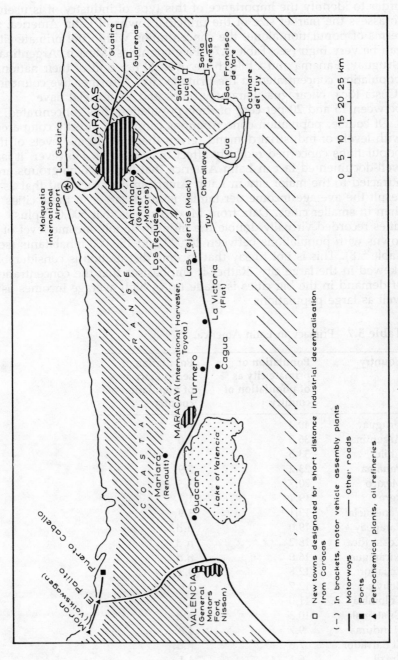

Fig. 5.3 The Venezuelan axial belt *Source*: Gwynne (1985)

At the same time, the major metropolitan city provides the optimal location for industries whose major market is that city. In order to identify the importance of this type of industry, it is useful to assess the market size of the principal towns of Latin America. In terms of population alone, the concentration within the primate city can be very high (see Table 5.7). Five countries (Mexico, Argentina, Uruguay, Panama and Chile) have over 20 per cent of their national population concentrated in the primate city. A further five countries (Costa Rica, Nicaragua, Paraguay, Peru and Venezuela) have between 15 and 20 per cent of their population thus concentrated.

Of course, population concentrations cannot be directly compared with levels of industrial demand, given that the income levels of the populations concerned radically affect that demand. However, it is a well-documented fact in Latin America that high-income groups are attracted to the major city in Latin American countries and that as a result the average income per family is much larger in these cities than in smaller cities. Data from Mexico shows that 500,000 plus cities recorded virtually double the monthly family income level of towns with populations between 10,000 and 150,000 inhabitants (see Table 5.8). This is not to say that income distribution is considerably skewed in the large city. Rather it emphasises that the concentration of demand in the big cities is made up of high average incomes as well as large populations.

Table 5.7 Primacy in Latin America, 1979

Country	Population of primate city as % of population of country	Population of second city as % of population of country
Uruguay	40.4	n.d.
Argentina	36.5	3.0
Chile	33.5	5.6
Panama	23.2	7.9
Mexico	20.7	3.5
Peru	19.1	1.7
Venezuela	19.1	5.9
Paraguay	19.0	n.d.
Costa Rica	18.2	n.d.
Nicaragua	16.0	n.d.
Ecuador	12.8	9.2
Bolivia	12.7	4.6
Colombia	11.0	4.4
Guatemala	10.3	n.d.
Honduras	9.7	4.8
El Salvador	7.6	n.d.
Brazil	6.1	4.1

Source: Gwynne (1982)

Table 5.8 Size of city and income levels in Mexico

Size of city (inhabitants)	Monthly family income (pesos)
Less than 2,500 (rural)	740
2,501–10,000	1,000
10,001–150,000	1,450
150,001–500,000	1,900
500,000 plus	2,800

Source: Ternent (1976)

It could be argued that, although incomes are higher in the big cities, costs are also higher – a factor that would serve to reduce industrial demand. Unfortunately there are very little data on the relationship between city size and costs in Latin America. It would appear that housing costs are higher in the bigger cities as are, due to greater journey-to-work distances, transport costs. However, an interesting retail survey at the end of the 1960s in Chile demonstrated that retail costs were not necessarily higher in the big cities than in cities further down the urban hierarchy. Indeed the relationship between city size and the retail price index could be described as an inverted U-shaped curve, with the lowest average price indices occuring at both ends of the city-size range. Although such evidence is only partial to the overall cost structure of large cities, it does imply that the higher relative incomes enjoyed in the larger cities are not necessarily reduced in comparative terms by higher costs.

Cheaper labour costs at other locations could attract industrial entrepreneurs away from the primate city. It has often been assumed that labour costs in the primate city are higher than the national average and considerably in excess of those prevalent in the peripheries of Latin American countries. However, a comparison of the Chilean 1967 Industrial Census (all firms over five employees) and the 1978 Industrial Survey (all firms over fifty employees) reveals a different spatial distribution and structure of labour costs (see Table 5.9). In terms of both white- and blue-collar workers, the Chilean primate city of Santiago had both average salaries and wage rates below or near the national average on both occasions. It should be borne in mind that between 1967 and 1978, Chilean industry underwent structural changes, first undergoing Marxist policies of state control and then experiencing monetarist policies stressing comparative advantages. Despite such major shifts in industrial policy and direction during the period, the pattern of provincial ranking of industrial salaries and wages remained remarkably stable.

Those provinces dominated by export industries tended to have salary and wage rates higher than that of the primate city. The average ranking of the copper-exporting provinces of O'Higgins,

Table 5.9 The changing rank of Chilean Provinces between 1967 and 1978 in terms of salaries and wages

Province	Ranking in terms of provincial average for white-collar salaries in manufacturing		Ranking in terms of provincial average for blue-collar wages in manufacturing	
	1967	1978	1967	1978
Tarapacá	9	4	6	3
Antofagasta	6	5	2	6
Atacama	1	12	1	
Coquimbo	19	16	9	18
Aconcagua	17	15	14	13
Valparaíso	5	6	4	5
Santiago	7	7	10	8
O'Higgins	2	1	7	1
Colchagua	10	8	11	7
Curicó	22	14	21	20
Talca	18	10	13	9
Linares	8	21	15	12
Ñuble	15	13	19	10
Concepción	4	9	3	4
Arauco	21	2	24	14
Bío-Bío	3	3	5	15
Malleco	13	17	20	21
Cautín	20	22	18	22
Valdivia	16	18	17	16
Osorno	11	11	12	11
Llanquihue	14	19	16	17
Chiloé	23	24	22	24
Aysén	24	23	23	23
Magallanes	12	20	8	19

Source: Gwynne (1985)

Atacama, and Antofagasta in terms of both white- and blue-collar workers was high in both years, while the timber- and cellulose-exporting provinces of Bío-Bío, Arauco and Ñuble improved their combined average ranking substantially during the period (from thirteen to six in terms of white-collar workers). As export industries are linked to world rather than national markets, there is greater elasticity in wage and salary rates in regions dominated by such industries. However, more significant for the location of consumer goods industries is the fact that Chile's second and third industrial conurbations of Valparaíso and Concepción generally had higher average salaries and wages in manufacturing than Santiago on both occasions. The major industrial decentralization schemes of Chile, located in Arica and Iquique in the province of Tarapacá, also had higher salaries and wages than the primate city. Thus, in Chile, the most attractive alternative locations for industrial entrepreneurs had higher average labour costs than those prevalent in Santiago. Meanwhile, Santiago could offer a wider range of labour skills and training than these alternative locations.

Industrial entrepreneurs may choose a metropolitan location due to what are known as the economies of agglomeration. Agglomeration economies are generally divided into economies of localization and economies of urbanization. Localization economies are those gained by firms in a single industry (or set of closely related industries) at a single location, economies accruing to the individual production units through the overall enlarged output of the industry as a whole at that location. Urbanization economies apply to all firms in all industries at a single location and represent those external economies passed on to enterprises as a result of savings from the large-scale operation of the agglomeration as a whole.

Localization economies may at first glance appear to have limited applicability in large Latin American cities. However, there is growing evidence of the spatial concentration of some industries in certain cities – such as the Brazilian motor vehicle industry in the Greater São Paulo conurbation. Furthermore, there is growing evidence of subcontracting linkages within industries in large cities. Complex networks of production have evolved in which a modern factory may commission a small family enterprise to undertake part of its production such as in the assembly of refrigerators, the upholstery of buses or dressmaking. Even when the modern factory is located in a peripheral region, such subcontracting linkages may be concentrated in the primate city – as with the early development of the Chilean motor vehicle industry. In 1968, when the majority of Chilean motor vehicles were assembled in the northern port of Arica and when over 50 per cent of parts by value had to be Chilean-made, 89 per cent of the 300 parts-producing plants that had been created in Chile were nevertheless located in Santiago; only 8 per cent of the plants had decided to locate near the assembly plants in Arica. The metalworking and other small firms that ventured into components production in Santiago were generally not new firms but existing companies that were intent on adding a further line of production to their organization. The concentration of this small-scale manufacturing economy in the primate city is one of the principal elements of localization economies in Latin America and can effectively reduce costs for the large-scale firm located there.

Urbanization economies can also have powerful attractions for industrial decision-makers. The effect of economies of massed reserves and large-scale purchasing is to reduce the per unit cost of inputs and reduce the amounts of non-productive capital in the production process in comparison with similar cities. Furthermore, the nature of urban systems in Latin America provides the primate city with a greater variety and quality of services. An example of this is banking. In Chile, all major financial decisions made by the provincial branches have to be referred to the bank's head office in Santiago for ratification; entrepreneurs in the capital will therefore have a much more rapid and often favourable access to capital than those in the provinces. Similar concentration of banking functions are found in Mexico where the Federal District accounts for 68 per cent of the total capital stock and reserves in the national banking system and 93 per cent of the long-term deposits. Furthermore, the banks of the Federal District grant 76 per cent of the national total of mortgage loans and handle 68 per cent of the country's investments in stocks and bonds.

Finally the infrastructure of primate cities, with an international airport, central railway system and spare power capacity has many advantages over the comparable infrastructure of smaller cities. In many Latin American countries, there have been cases of small towns lacking the necessary power capacity to cope with the operation of new high-energy industries at peak times; the consequent stoppages can have severe cost penalties for industries using continuous casting or other such processes. An example from Chile is interesting in this respect. In 1969, the Chilean company, Manufacturas de Cobre (MADECO), decided to locate a plant making copper telephone cables in the northern port of Antofagasta. Despite the fact that Antofagasta was the fourth city of Chile and despite its proximity to the Chuquicamata copper mine, an enormous user of electricity, the MADECO plant suffered for seven years from cuts in electricity supply at peak periods of demand. As MADECO had smelters which used a continuous casting process, where temperature is kept at a constant to produce the required type of refined copper, a cut in electricity meant the total loss of copper being refined and a period of no production as the smelters cooled, were cleaned out and prepared again for production.

However, it has frequently been pointed out that the considerable infrastructure that the capital city possesses has normally been heavily subsidized by national funds and that the entrepreneur in the primate city does not pay the real costs for such services. For example, in Mexico, the price of electricity is the same in Mexico City as in all other parts of the country despite the fact that the bulk of electricity consumed by the city is generated at the hydroelectric complexes of Malpase and Infiernillo, 1,000 and 600 km (621 and 373 miles) away respectively. The considerable cost of transmission is not directly passed to the entrepreneurial consumer.

Processes of industrial decentralization

The large primate city does, however, have some disadvantages for the location of industry in Latin America. The cost of land is normally higher and tends to dissuade land-intensive industries. The increasing congestion of many large Latin American cities makes

Plate 9 Flatted factories in Caracas. Due to intense pressure on land in the area, new industrial building takes the form of flats. Pharmaceutical, optical and office equipment industries are attracted to such locations. (Photo: R. N. Gwynne)

transportation more difficult both for people and products. The costs of housing, health, recreation and labour *can* also be higher in the major city.

The potentially high costs of locating or even maintaining an industry in the primate city can be found in the Venezuelan capital, Caracas. In 1976, the Caracas agglomeration had a population of 2.6 million, crowded into an elongated but narrow basin whose east–west axis measures 16 km (10 miles) from Catia to Petare but whose north–south dimensions rarely exceeded 3 km (2 miles). To the north, the steep Cordillera de la Costa prevents any expansion and indeed the area has been designated as a national park. To the east and west, high land similarly excludes development. It is only in the south where five valleys penetrate hills of more moderate but still significant elevation that outward expansion of the city can still occur – an area that because of high costs of construction, land and transport is becoming an exclusive residential zone for the middle classes.

The unavailability of flat land for industry after 1966, the concomitant high land prices and congestion have acted as powerful decentralizing forces for Venezuelan manufacturing. In no sector is this better represented than in the motor vehicle assembly industry where the use of large areas of cheap flat land is at a premium. As

131

Fig. 5.3 demonstrates, nine vehicle assembly plants are now located outside Caracas in the axial belt stretching from Las Tejerías to Valencia and Morón. Only the original General Motors plant remains in the Caracas agglomeration, in the southern industrial suburb of Antimano.

The movement of assembly plants into the axial belt effectively began with the introduction of Ford as a new producer in 1963 and that company's decision to locate its new plant at a green field site on the outskirts of Valencia. Within four years Ford had become the major vehicle manufacturer of Venezuela, overtaking the two major US producers, Chrysler and General Motors, in the process. Chrysler and General Motors found it difficult to react to Ford's increases in market share, partly because of the restrictions their plant site put on productive expansion. This was felt most acutely by Chrysler which had a site near central Caracas, in which further expansion was impossible. As a result, Chrysler decided to sell its plant in Caracas, followed Ford's locational decision and located a new plant on a site in Valencia with ample room for expansion. The selling of the Caracas plant almost covered the costs of the land and new plant in Valencia. As a result, the production of Chrysler vehicles was able to expand and by 1973 Chrysler had pushed General Motors into third place in terms of production. This position lasted until 1979 when General Motors bought out the Chrysler plant and started producing its own vehicles there. Since then, General Motors has been concentrating production at the ex-Chrysler plant in Valencia and using the Antimano plant for more specialized activities.

Industrial decentralization has therefore been occurring in Venezuela but in the form of movement along the Caracas-Valencia axial belt. The well-established towns of Valencia and Maracay have been the main recipients of new and relocating industrial plants. Small towns, however, such as La Victoria, and even large villages, such as Mariara, have attracted significant industrial enterprises. Meanwhile, government planners have been intent on promoting the short-distance movement of industry out of Caracas to a collection of new towns in the middle Tuy valley and a large planned town to the east, Ciudad Fajardo, that will be a bi-nodal agglomeration combining the settlements of Guarenas and Guatiré (see Fig. 5.3). Nevertheless, in 1974, 47 per cent of Venezuela's 300,000 manufacturing workers were employed in the Caracas agglomeration (including the port of La Guaira). Increasing numbers were working in flatted factories such as those in La Urbina built in the far east of the Caracas basin near Petare. Meanwhile, only 90,000 manufacturing workers (30 per cent of the total) were employed in the other settlements of Venezuela's axial belt in 1974.

This process of industrial decentralization from Caracas into the axial belt is being encouraged further outwards by the Industrial Decentralization Policy of 1974. Few benefits can be achieved by any

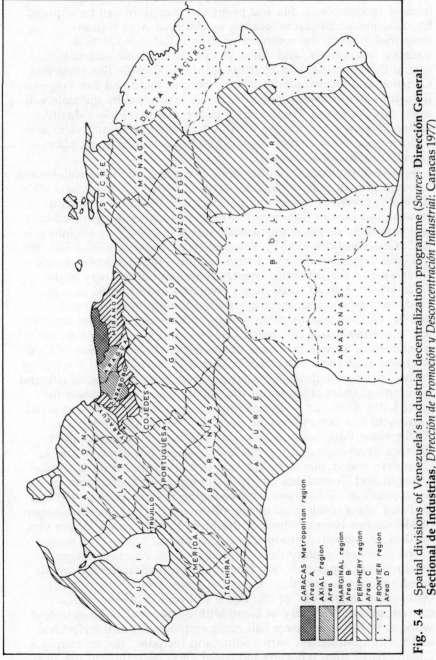

Fig. 5.4 Spatial divisions of Venezuela's industrial decentralization programme (*Source:* **Dirección General Sectional de Industrias,** *Dirección de Promoción y Desconcentración Industrial:* Caracas 1977)

but high priority industry in Caracas (Area A) or the Caracas – Valencia – Puerto Cabello axis (Area B) (see Fig. 5.4). Benefits in the form of favourable credits and profit tax reductions can be acquired for most industrial plants locating in Marginal Area B (non-industrial areas to the south of the Puerto Cabello–Valencia–Caracas–Guatiré axis) and Areas C and D. The basic idea is to extend the industrial belt both eastwards (to include Barcelona and Cumaná) and westwards (to include Barquisimeto and San Felipe). Industries whose development is closely controlled by the state will assist in the process. For example, in the motor vehicle industry, General Motor's new engine plant will be built at Barquisimeto and a Fiat engine plant at Barcelona, while Pegaso's new truck plant is already producing at Cumaná.

Such decentralization policies seek to expand upon the well-known resource growth pole of Ciudad Guayana in eastern Venezuela. This famous experiment in regional development based on abundant mineral and energy resources has been planned and organized by a semi-autonomous government body (the CVG) since its beginning in 1958. Generously funded by Venezuelan oil money in the 1970s, the CVG has created an iron and steel complex, massive hydroelectric potential and the first fully integrated aluminium industry in the Third World with all three stages present – extraction of bauxite, production of alumina and the refining of aluminium. A small number of steel and aluminium-using industries have been attracted to the frontier towns of Ciudad Guayana and Ciudad Bolivar, but it is interesting to note that most of the national consumption of CVG steel and aluminium is located in Venezuela's axial belt.

Short-distance industrial decentralization from the primate city and government encouragement of the phenomenon are common to other Latin American countries. In Mexico, short-distance industrial movement has been prevalent along the Mexico City–Puebla axis. At the same time, Mexican governments have encouraged short-distance decentralization by establishing large industrial estates at Queretaro and Irapuato, by creating new industrial cities at Ciudad Sahagún and Cuernavaca and by planning an industrial zone of the west, south of Guadalajara, and including the towns of Octlán and La Barca. As a result most of the effective industrial decentralization of Mexico has been confined to the Mesa Central where 70 per cent of Mexican industrial production is generated.

The twentieth-century development of Latin America's major industrial agglomeration, São Paulo, has been characterized by the steady onion-like growth of industrial zones around the original centre. The 'municipio' of São Paulo still represents the largest concentration of industry in Brazil with 24 per cent of all industrial workers in Brazil and over half of all employment in the electrical and plastics industries. Surrounding and including the municipio is Greater São Paulo, reflecting industrial decentralization in the 1950s and 1960s when the ABC towns of Santo André, São Bernardo do

Campo and São Caetano do Sul became attractive for a wide range of industry but particularly that of motor vehicles. The vehicle plants of Chrysler, Ford, Mercedes Benz, Saab-Scania, Toyota and Volkswagen are concentrated in Brazil's vehicle capital of São Bernardo with the General Motors' plant in neighbouring São Caetano. In the 1960s and 1970s, however, new industries have located in and older industries have moved out to the smaller inland towns of São Paulo state. It is these towns of between 200,000 and 500,000 inhabitants where industrial productivity is highest in terms of urban size. São Paulo's industrial engine has moved out along three axes (see Fig. 5.5): to the north-east along the Paraiba valley to incorporate the towns of São José dos Campos and Taubaté, where in particular new component and vehicle plants are locating; to the north-west along the Paulista railway line to incorporate such towns as Campinas, Piracicaba and Americana, where engineering and textiles stand out in a wide range of industry; to the west to incorporate such towns as Sorocaba, where non-metallic minerals, food processing and textiles are the major sectors.

The spatial association of industrialization with the primate city

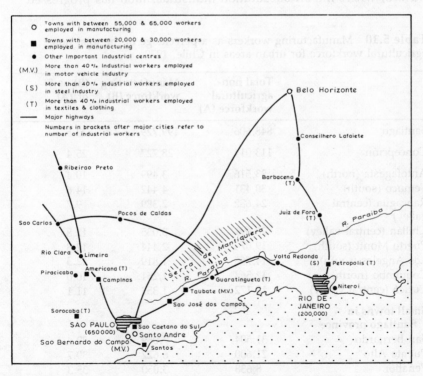

Fig. 5.5 The industrial core of Brazil *Source*: Gwynne (1985)

has had the corollary that towns further down the national urban hierarchy and distant from the primate city have attracted little industry. In particular, the inward-looking, consumer-good industries were strongly attracted to the large city and avoided the smaller provincial towns. The need to be near the major internal market and to have easy contacts with foreign technology and companies were always vital considerations for these industries. As a result, manufacturing has had relatively little impact on the occupational structures of small- and medium-sized towns distant from the primate city. Table 5.10 demonstrates this relationship in the Chilean urban hierarchy. Santiago and the port-industrial town of Concepción had approximately one-quarter of their substantial work forces in the manufacturing sector. Medium- and small-sized towns away from Santiago had only about one in eight of their workers employed in manufacturing. The attraction of industry for a location near the primate city can be gauged from the differential performance of medium- and small-sized towns located near Santiago. San Bernardo, Puente Alto and Peñaflor had up to 35 per cent of their work force involved in manufacturing (see Fig. 5.6).

Such short-distance decentralization has become most associated with the larger more dynamic industrial countries of Brazil and Mexico, where import substitution industrialization has progressed

Table 5.10 Manufacturing workers as a percentage of the total non-agricultural workforce for urban areas in Chile, 1970

	Total non-agricultural workforce (A)	Manufacturing workforce (B)	(B) as % of (A)
Santiago	848,606	199,729	23.5
Concepción	113,014	28,727	25.4
Antofagasta (north)	33,516	3,499	10.4
Temuco (south)	30,431	4,442	14.6
Rancagua (central valley)	24,682	2,389	9.7
Chillán (central valley)	23,691	3,035	12.8
Puerto Montt (south)	19,074	2,444	12.8
Los Angeles (south)	14,765	1,819	12.3
Coquimbo (north)	12,564	1,251	10.0
Curicó (central valley)	12,172	1,353	11.1
Small towns in Santiago province			
San Bernardo	31,504	7,069	22.4
Puente Alto	21,600	6,622	30.7
Peñaflor	8,638	3,050	35.3

Source: Instituto Nacional de Estadisticas (1973)

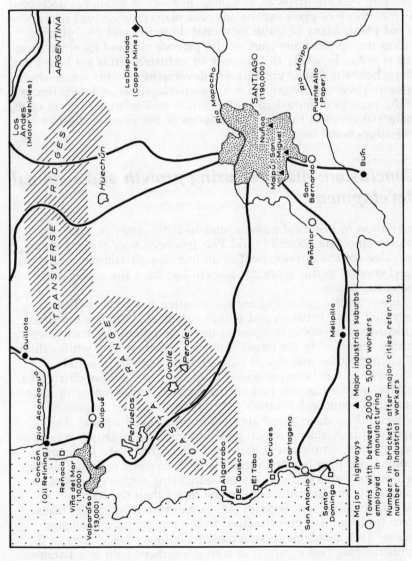

Fig. 5.6 Chile's central region *Source:* Gwynne (1985)

furthest and has created a highly diversified, closely interlinked urban-industrial structure. Within a national framework of rapid economic growth, both multinational and domestic industrial firms aim to continually expand production and diversify into new, often related, areas. Successful multinational and domestic firms can therefore become involved in a large number of locational decisions, such as those of plant expansion, new plant creation and relocation of old plant. Many of these locational decisions will take place within the spatial framework of the primate city and its environs as this is not only where the majority of industrial firms are based but also where the fastest growing are developing. In this sense, short-distance decentralization can best be envisaged as an intensification of the process of industrial centralization and as an expansion of the industrial core out from the boundaries of the primate city to a wider functional region beyond.

Conclusion: manufacturing growth and regional development

In relation to historical growth rates in Latin America, economic growth since the Second World War has been very rapid and has provided for the development of an increasingly complex society. Fundamental to this economic growth has been the process of industrialization.

However, the process of industrialization has been spatially restricted, both between and within countries. Complex industrial structures comparable to those of developed countries have been mainly restricted to the larger countries. Furthermore, within almost all countries, the majority of manufacturing production and employment has been generated from the major metropolitan area.

What impact has this had on the regional development of Latin American countries? Invariably it has led to greater disparities between the industrialized and less industrialized regions. The rapidly industrializing region generally provides steady increases in employment, employment that has both higher wage levels and productivity than labour in the less industrialized region. This has two primary effects. First, the industrializing region continues to increase its productivity and income faster than the less industrialized region. Second, substantial labour migration takes place from the less industrialized to the rapidly industrializing region. This maintains the internationally lower wage rates of the industrializing region which in turn encourages both international and national holders of capital and entrepreneurs to start or continue investing in the region. In this way a cumulative process of increasing regional inequality can be set in motion.

In the industrially more dynamic metropolitan areas such as São Paulo, the growth in the labour market is broadly matched by growth in both secondary and tertiary employment. In industrially less dynamic agglomerations such as those of Lima and Santiago, large gaps can develop between labour supply and demand. Large-scale unemployment and underemployment result. The underemployed attempt to survive in an economy of petty trading and petty services. In these cities the workers regularly employed in manufacturing tend to form a labour aristocracy, with better wages, conditions of work and security than those workers outside the sector.

Industrialization has not been a panacea for all of Latin America but where it has concentrated, as in São Paulo state, it has produced regional economies as prosperous and dynamic as those of Western Europe. Already the per capita income of São Paulo state is equivalent to that of a European country such as Portugal. Where manufacturing growth has been more modest, employment generation has fallen far short of the labour force attracted. In the many peripheral areas of Latin America, away from the cities, mines and plantations, modern manufacturing is still largely absent.

Further reading

Cunningham S M (1981) 'Multinational enterprise in Brazil: locational patterns and implications for regional development', *Professional Geographer* **33** (1), 48. (Good introduction to the subject.)

Dickenson J P (1978) *Brazil*. Dawson, Folkestone. (A useful reference book for the sectoral and spatial development of industry in Brazil.)

Ffrench-Davis R and **Tironi E** (eds.) (1982) *Latin America and the New International Economic Order*. Macmillan, London. (External constraints on economic and industrial development in Latin America are analyzed comprehensively.)

Gereffi G and **Evans P** (1981) 'Transnational corporations, dependent development and State policy in the semiperiphery: a comparison of Brazil and Mexico', *Latin American Research Review* **16** (3), 31–64. (Good introduction to the subject.)

Gilbert A (1974) 'Industrial location theory: its relevance to an industrialising nation', in **Hoyle B S** (ed.) *Spatial Aspects of Development*, pp. 271–290. John Wiley, London. (Spatial concentration of Latin American industry in large cities is explained.)

Gwynne R N (1978) 'The motor vehicle industry in Latin America', *Bank of London and South America Review* **12** (9), 462–471. (An introduction to the rapidly expanding motor vehicle industry in Latin America.)

Gwynne R N (1982) 'Location theory and the centralization of industry in Latin America', *Tijdschridt voor Economische en Sociale Geografie* **73** (2), 80–83. (Spatial concentration of Latin American industry in large cities is explained.)

Gwynne R N (1983) 'When trade stops being the engine growth', *Geographical Magazine* **55** (10), 503–507. (Details how the 1979–83 world recession affected the industrial economies of Latin America. A series beginning with this article.)

Gwynne R N (1985) *Industrialisation and Urbanisation in Latin America.* Croom Helm, London. (The author's recently published book on topics covered in this chapter.)

Hirschman A O (1968) 'The political economy of import-substituting industrialization in Latin American countries' *Quarterly Journal of Economics* **82**, 1–32. (The theory of industrialization through import substitution is well synthesized in this article.)

Instituto Nacional de Estadisticas (1973) *Caracteristicas Basicas de la Población, Censo 1970.* Santiago.

Morley S A (1982) *Labour Markets and Inequitable Growth: the Case of Authoritarian Capitalism in Brazil.* CUP, Cambridge. (Assessment is made of the impact that industrialization has made on regional labour markets.)

Smith C H (1980) *Japanese Technology Transfer to Brazil.* UMI Research Press, Ann Arbor, Michigan. (A clearly presented and informative case study on the transfer of technology.)

Ternent J A S (1976) 'Urban concentration and dispersal: urban policies in Latin America, in **Gilbert A** (ed.) *Development Planning and Spatial Structure.* John Wiley, New York.

Tussie D (ed.) (1983) *Latin America in the World Economy.* Gower, Aldershot.

CHAPTER 6

Regional development, trends and policies

Arthur Morris

This chapter takes four different themes in order. The first is the variation in level of income between and within the Latin American Republics. From this starting point, consideration goes on to the theories of regional development, especially those developed in, or applied to, Latin America. Several regional planning types are then studied and related to theory and to the initial problem of varying income levels. It is shown that planning has been poorly executed and is often at odds with real regional needs. In view of the more recent European trends, some reorientation of attention is suggested in a final section on regionalism.

Income inequalities

One of the most striking aspects of Latin America, which impresses the visitor immediately he descends from an aeroplane at any of the major airports, is the wide gap between rich and poor. The startling juxtaposition of shanty town and wealthy suburb with which he is confronted is only the most obvious example of the gap. Others, less evident but equally real, are the inter-regional contrasts between metropolitan central regions and rural peripheries, and the international differences of wealth and levels of living.

We may begin with the international differences of per capita production (Gross Domestic Product or GDP data are the most accessible and complete series). In 1980, per capita GDP ranged from 2,470 dollars in Venezuela, to 275 in Haiti, a little less than a 10 to 1 ratio. This range vastly exceeds Western Europe's where there is a 3.65 to 1 ratio between Switzerland and Greece, and is comparable to the range in Africa south of the Sahara, between the extremes of South Africa and Burundi (10.1 to 1). The range as an index is unsatisfactory as it measures only the extremes, and a more complete measure using each country and weighting values by population size, is the Coefficient of Variation (CV). For 1980, this index was 39.8 per cent for Latin America, a high level of variability

141

and showing a major increase over the 1960 value which was 27.9 per cent.

This increase has received little attention by students of Latin America and some preliminary comments may be in order here, though the theories mentioned later in the chapter will also provide some explanation. Two points may be made: first, there is no evidence that the Latin American internation groupings, such as LAFTA, CARICOM, or the Andean Group, have done anything to reduce the economic differences between their members, though they have been in existence for more than a decade. This may be attributed both to their weakness of structure as institutions, and to the continuing dominant influence of north-south relations. Secondly, there would seem to be something of a core-periphery structure on an international, Latin American scale, which would also account for some of the differences. Argentina, and particularly the Buenos Aires region, lies at the heart of an economic system involving Chile, Bolivia, Paraguay and Uruguay. Venezuela is central to another system involving Colombia, Ecuador and parts of the Caribbean. These systems are recorded indirectly in the GDP differences, but they also involve movements of skilled workers, of capital investment and savings, focussing on the centres, as has been shown elsewhere (Morris, 1981). The opening up of major road, rail and air links between the countries, replacing in importance the old links to port cities and overseas markets, is reinforcing this new centre-periphery structure.

It is of course possible to measure the level of living and the rate of development in other ways than through use of GDP per capita. Income from production is not necessarily spent on the welfare of the population, or even spent locally at all. Much income is not to be measured in monetary terms at all – it is for example, difficult to place a value on the relative incidence of air and water pollution. Those items for which data are available tend to be correlated with production and income variables; thus a widely available statistic is the population divided by the number of hospital beds. Here the variation from country to country is comparable to that of GDP (Haiti 1,239 persons per bed, Argentina 176, ratio 7 : 1). In Western Europe the ratio is only 2.8 : 1 and the worst provision is in Portugal with 187 persons per bed; while in Africa the provision is poorer and the extremes wider (Ethiopia 3,016 persons per bed, Equatorial Guinea 95 persons).

Turning to the regional, within-nation disparity levels, the evidence is inconclusive. Some of the results for different countries, using varying time periods and data, are collated by Gilbert (1976 and 1978). Declines of regional inequality in Brazil and Colombia are balanced by increases in Argentina, Chile and Venezuela in the studies he cites, so that it is difficult to discern overall trends. It is instructive to examine in detail the figures for one country. Those for Mexico, for 1970 and 1980, are presented in Table 6.1 (see also Fig. 6.1). In this case the CV Index shows a strong decrease from

Table 6.1   Inter-regional variability of GDP per capita, Mexico 1970 and 1980

State	1970 GDP per cap. (Mex. Pesos)	Percentage Contribution to index*	1980 GDP per cap. (Mex. Pesos)	Percentage Contribution to index*
Aguascalientes	6,095.8	0.15	51,629	0.13
Baja California	13,483.0	1.24	78,225	0.61
B. California Sur	14,550.0		81,317	0.13
		0.31		
Campeche	6,626.4	0.07	61,149	—
Coahuila	14,193.3	2.11	81,084	0.83
Colima	12,759.1	0.25	68,505	0.02
Chiapas	3,358.1	3.03	53,861	0.37
Chihuahua	8,099.9	0.04	62,279	—
Dis. Federal	20,711.3	63.76	114,692	49.87
Durango	5,431.4	0.66	51,407	0.33
Guanajuato	5,123.5	1.96	43,516	2.31
Guerrero	5,589.9	1.05	33,092	3.91
Hidalgo	4,416.5	1.49	45,391	0.91
Jalisco	6,718.7	0.90	64,962	0.02
Mexico	7,521.1	0.39	55,489	0.96
Michoacán	4,013.2	3.41	34,449	4.82
Morelos	6,384.5	0.22	56,179	0.10
Nayarit	6,012.1	0.26	43,135	0.60
Nuevo León	15,254.8	4.60	101,803	7.62
Oaxaca	3,107.6	4.23	23,717	7.36
Puebla	4,856.1	2.50	38,588	4.20
Queretaro	5,604.9	0.32	54,749	0.10
Quintana Roo	6,611.8	0.03	67,799	—
San Luis Potosí	6,133.7	—	33,413	2.76
Sinaloa	6,774.5	0.57	54,912	0.27
Sonora	12,474.4	0.98	69,987	0.13
Tabasco	4,056.5	1.12	109,078	4.98
Tamaulipas	8,819.7	—	77,332	0.79
Tlaxcala	2,992.1	0.94	35,289	0.84
Veracruz	6,546.0	1.23	49,911	1.93
Yucatán	6,271.6	0.31	49,243	0.38
Zacatecas	3,381.2	1.82	27,762	2.71
MEXICO	8,777.6	100	63,466	100
Overall index		64.5		42.9

Source: Wilkie and Haber (1983)

*The index used is the Coefficient of Variation (CV) $= \sqrt{\dfrac{\Sigma(y_i - \bar{y})^2 \cdot n_i/n}{\bar{y}}}$

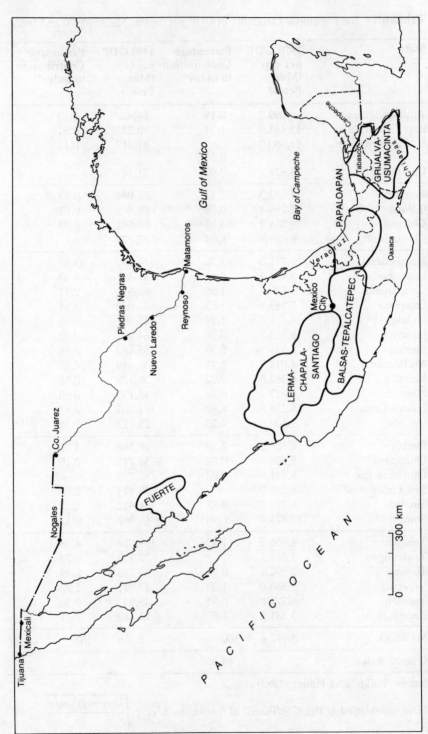

Fig. 6.1 Regional development in Mexico. A strong geographical pattern existed in the 1960s and 1970s, the US border being the location for assembly industries and the centre-south the main focus of river basin development schemes (based on Barkin and King 1970)

1970 to 1980, from 64.5 per cent to 42.9 per cent. This falls in line with other findings (e.g. Ginneken 1980), which show a divergence trend up to 1960 and a fall thereafter to 1970. To interpret these findings it is however necessary to dissect the index and identify the contribution of individual states, as is done in Table 6.1, where columns 3 and 5 show the percentage contribution of each state to the overall index value. The Federal District of Mexico City evidently dominates the scene, with two-thirds of the total in 1970 and half in 1980. This decline of Mexico City over the decade is in fact what is largely responsible for national decline in interstate variation of GDP, so that the index is not really instructive about other regions. Another rather unreal effect is the result of Mexico City's expansion out into the surrounding Mexico State; this has meant an effective spreading of metropolitan wealth. Apart from this, the decline of metropolitan wealth may be seen as due to massive immigration of poor people from states into Mexico City. Elsewhere in the country, polarization of income and GDP would appear to be continuing through the 1970s, concentrating in such regions as the southern Gulf oil area, represented by Tabasco.

Mexico in the 1970s may be seen as a country with strong economic growth canalized into a limited number of zones, but exhibiting something of a levelling process in the industrial and service sectors, though agriculture continues to be a source of differentiation. Industry has been encouraged to develop in the northern border towns, which were given special tax status to attract US-owned industries between 1965 and 1972, and some further industrial growth outside Mexico City has been fostered since 1972 by tax and other financial advantages. The services sector has seen a levelling process of another kind, produced by massive migration into Mexico City and lowering the relative wage rates there. Agriculture has had growing differences between the regions, because high-technology irrigated agriculture has been applied in the north on large commercial farms, while the centre and south have remained traditional, poor farming areas.

The differences between sectors in terms of their effect on interregional variations are illustrated by Gini-coefficients, used by Ginneken (1980) for 1950–70. Overall, the last column shows a

Table 6.2 Gini-index of concentration of income by sector over ten major regions of Mexico 1950–1970

Year	Agriculture	Sector Industry	Services	Total
1950	0.24	0.21	0.21	0.25
1960	0.29	0.19	0.20	0.29
1970	0.36	0.20	0.21	0.25

Source: after Ginneken (1980)

145

modest reduction in 1960–70, but agriculture's widening gap prevents the reduction from being greater.

Regional development concepts

From the Second World War to about 1970, a dominant line of thought in social science was that a single path of development was likely from poor to rich, that it was paralleled by a transition from traditional to modern, from rural to urban and from agricultural to industrial. In spatial terms this multi-faceted process of development would have expression as an initial rise in inter-regional differences, i.e. divergence, as early development took place in one or two favoured localities, followed by a spread of development to all other areas, and in consequence a decline of inter-regional differences, or convergence, to very low levels, representing some residual friction in movement of the factors of production.

The concept of spread was an important aspect of the theory. Mechanisms by which the spread of economic development could take place were to be found in neo-classical economics, which predicted factor movements in an economy not bounded by space friction, so that factor returns would be equalized over the regions. In simple terms, this meant that workers would migrate from low-wage peripheral regions to high-wage central regions, and in so doing would reduce the differences in wage rates between the regions; while capital would migrate from central regions of low returns, to peripheral areas where returns were higher, and thus even out the differences in returns to this factor.

The criticisms of these views are of a diverse nature, and different lines of thought may be traced out. Some points are, however, accepted in common and may be noted at once. Firstly, factor movements do not take place in the simple fashion predicted. There are physical, informational and perception barriers to the movement of both people and capital in a country. Migration moves through channels formed by community bonds and family ties, which only distantly relate to wage differences. Capital movement does not take place as a flow at all because it consists of large individual investments in factories and equipment. In any case, capital is attracted to central, metropolitan regions by powerful forces – the economies of scale available through producing for a large market, and the external economies available from linking to associated industries in the centre. For both capital and labour, the costs or difficulties in transport and communications between regions impose serious obstacles to spread.

Beyond these strictly economic factors, most commentators would agree that 'structural' factors are also important. These factors refer to the varied institutional and social structures within which the economy must operate, and vary from the effects of fiscal legislation, to the operation of international trade and the structure of land holdings in agriculture. Import tariffs in Latin America have often favoured consumer goods manufacture over capital or intermediate goods, which meant concentration in the centre where consumer markets were established; internally, raw materials exports from poor regions were taxed while personal incomes, highest in the centre, were little affected by tax. International trade, it has been argued, has offered generally poor terms of trade to raw materials producers, and the same applies to food/raw material-producing regions in one country. Another structural factor mentioned is the ownership of rural, and sometimes also urban land, where there is a strong bias, a very small percentage of owners controlling most productive lands and the major urban businesses and properties.

Beyond such points of general agreement, the arguments vary, in interpretations and in proposals, and indeed in their fundamental orientation. We may consider two important lines of thinking here: development from below, and dependency theory.

Development from below

Development from below has a diverse set of components, as it has incorporated much of the criticism of the diffusionist paradigm. It uses Michael Lipton's criticism of the urban bias of existing development, Ernest Schumacher's indictments of large-scale development and of outside technology, the ecologists' warnings against one-sided single-resource development and lack of attention to environmental questions, and the line of argument presented by the International Labour Office arguing for more attention to basic needs by the provision of work, food and shelter. It does not constitute a theory in the sense of the diffusionist arguments which predict precise factor movements and measure the forces inducing or holding them back. It is rather a loose-knit set of ideas, almost a change of viewpoint rather than a direct challenge to diffusionism. On the other hand, it is stated in the form of positive proposals and some of its elements are directly translatable into plans.

Ideas and proposals for development, according to this way of thinking, should originate from within a region or community, so that participation by the community would be guaranteed, and community involvement would itself ensure long-term success rather than the failure that often ensues when the (outside) promoters of current plans go away to start on another project. Coming from within, the developmental programme would use locally available human and material resources, not relying on outside technology and imported materials, again a firmer base for long-term success. Development would respond to local needs rather than national, because the ideas and projects chosen would be those selected by the participant population; participation in planning would be an inbuilt feature. Because of these characteristics, the programme would be diversified to meet many different needs, involving town and country in interlinked projects, and it would also tend to vary from one region to another because of the different tastes and interests in each region. One community might set much store by schools and colleges, another might prefer good roads for its farmers to reach their markets. To achieve these varied ends, selective regional closure is required, a closing-off of the region to the more negative effects of external influence, and promotion of internal relations instead. Overall, development from below ideas have much in common with those of nineteenth-century populism, in their emphasis on non-economic values and the need to provide for the whole man. They do not however have a strongly anti-industrial stance.

Dependency theory

The other critique of standard theory is the left-wing one, and the dependency theory version of it is the one which will be treated here as most relevant and certainly most applied to the Latin American case, since it was first developed by Latin American writers in the 1960s. The historical version of this is that the colonial territories of Portugal and Spain were subject to a powerful and crippling relationship of economic dependency on their mother country or metropolis. In opposition to the previous standard argument of spreading development, in which regions which were backward were so because of their isolation, or lack of effective diffusion of beneficial effects from the centre, the backwardness of regions was directly attributed to their relationship to the metropolis. It was indeed the strength of their central ties that caused their state of poverty, and dependency relationships might actually produce a negative development as for example, when previously local industries were forced out by competition with the centre.

The *peon* on a Chilean estate stood in a dependent relationship to the estate owner to whom he sold his labour; the owner was in turn dependent on traders who bought his crop of wheat and sold it in Santiago; in turn, Santiago merchants were dependent on those of Lima to whom the wheat was sold; and Lima merchants were dependent on Spain which controlled their export of silver and imports of manufactured goods. The argument is commonly presented as an indictment of capitalism, a fallacious argument since the dependency relationship existed precisely because there were no open markets on which each link in the chain could sell his product, but a single monopsonist. It is, however, a correct indictment of colonial mercantilism which was used by all the colonial powers to some extent. Through this system colonial powers could sell their manufactures to their colonies, and from them assure themselves of raw material supplies.

In modern times mercantilism is no longer possible, but it has been replaced by the increasingly strong role of the multinational corporations (MNCs). These are able to control the kind of operations engaged in, the locations chosen, the products of each of their overseas branches and are able to maintain an unequal exchange system as in colonial times by avoiding tax payment in the 'colony', paying low wages, making large transfers of profits to the metropolitan country, often hidden in various ways (see Ch. 5). States may also be effective in maintaining dependency, by limiting technology transfers, by their control of the terms of trade through stockpiling and trade agreements among themselves, and through the control of foreign aid.

It should be pointed out that small countries will tend to suffer particularly from dependency relationships, since they typically depend more on trade relations with the outside world than large countries, and because their range of products is smaller and more open to the negative effects of a decline in demand. Most Latin American countries are in fact 'small', not in the sense of population size or area, but in the size of their economy, the amount of foreign trade they do, and the range of products they export. A check of United Nations figures on these statistics will quickly confirm this small size, in comparison with the major economies of the rich countries.

If small countries are more exposed to dependency, by extension regions within a country should be still more so. Whether or not they in fact are more exposed depends on the kind of political system current in the country, its democracy or subjection to control by a ruling oligarchy or military group. It also depends on the cultural and ethnic structure found in the country. A specifically regional offshoot of dependency reasoning is internal colonialism. This concept has used the dependency ideas of the control of production and trade by the centre so as to exploit the periphery, and adds to the exploitation of one race or ethnic group by another, which preserves for itself urban employment, education and entry to administrative positions, consciously suppressing the dependent group's culture, languages, religion, and imposing its own. This concept was developed with respect to Mexico by Gonzalez-Casanova (1969) who observed, at a quite local level, the exploitation that went on from small country towns to their rural hinterlands. The former were the home of *mestizo* merchants and businessmen, who bought very cheaply the farm products of the peasants, while also selling them over-priced retail goods from the big cities and giving them credits at extortionate rates of interest. *Mestizos* controlled access to urban jobs and controlled the regional administration which was itself located in the town. Internal colonialism is extensible as a concept to all of Central America and the Andean countries, where there are substantial ethnic differences; but even dropping the racial element of exploitation, the dependency argument probably has force for Latin American regions where the town-country division is a central source of conflict.

The dependency critique and its variants lead to no positive suggestions for present policy. Reform movements to alter the present relations are usually attacked as 'reformism', and current attempts at regional development projects as legitimizing strategy, i.e. actions which will present the government in a favourable light to its electorate. Development from below on the other hand, accepts and works essentially within the existing framework of government, trying to restructure it in a gradual manner, rather than the revolution which is the overt or implied solution for dependency writers.

Application of the development from below ideas would itself require a substantial reorganization of government in some cases. The call for selective regional closure is the most difficult item to operationalize, and would require a well-minded and powerful national government to bring about, through the restriction of trade and communications, the promotion of local initiatives, and the acknowledgement of non-standard and non-comparable requirements in the different regions in such items as health and education, power supplies, housing and recreation. In this matter, the dependency inference of revolution might be accommodated by the devolution of power to regions as in some European countries, though such devolution does not necessarily mean any change in political or social structure.

One of the ways in which development from below has been operationalized is in the form of 'Agropolitan Development', in which a region of 50–150,000 population, with an urban centre of 10–25,000, will generate its own planning ideas and put them into action with its own technical and administrative skills (Lo and Salih 1981). Because of its democratic structure – the small size of region and of city ensures every inhabitant has access to administrators – this unit will tend to an equitable distribution of the fruits of any planning, supplying basic needs before all others, and basic needs as perceived by the constituent population.

If this kind of proposal cannot be applied it is likely to be because of the barriers the structuralists and dependency writers have described. First, central government may not really be in favour of regional policies at all, though it publicizes them as a legitimizing exercise. Secondly, regional policy may be sincere enough, but with the aim of helping national growth rather than that of the region. Thirdly, and possibly lying behind the first two, there may be a class or cultural difference which inherently prevents any real development; if government is exercised by a rich minority which seeks to preserve power for itself, it will not permit challenges from peripheral regions and their peoples. This factor may operate on a small scale too, between a small country where the dominant group lives, and the rural dominated one.

Regional development programmes

A view of regional development schemes based on the neo-classical ideas might be that inter-regional differences will be solved by migration of people and capital and that national investment should concentrate on the sectors and regions with highest returns; propping up declining or poor regions merely postpones their crises and perhaps aggravates them in the long term.

This cannot be regarded as a practical or useful approach for most countries. Apart from the criticisms raised already to show that market forces do not bring about the expected convergence, it is likely to be politically unacceptable for national governments to abandon peripheral regions to their poverty, under any other than a dictatorial regime. Instead, measures to alleviate the inter-regional differences must be seen as desirable. Those employed in Latin America, as will be shown in what follows, though they are ostensibly aimed at reducing differences of welfare and income, have often had quite distinct underlying motives.

There are various ways of classifying the different kinds of policies attempted in Latin America. Many policies have changed in character over the years so that they are intermediate in type. It is, however, useful to distinguish the broad-based policies from the narrow ones, and the policies aimed at place rather than people development. Policies such as the river basin approach, problem region approach, or integrated rural development, tend to be broad-based, while colonization, growth poles, and other similar urban-industrial policies are narrower-based. Regional development policies are 'place-development' when they focus on a particular area or territory, whether or not it is populated, or if populated, whether the occupants are particularly needy. River basin policies and colonization or frontier settlement policies, are of this kind, and so too on occasion may be growth poles. Integrated rural or rural-urban policies, and problem region approaches, are typically more 'people-development' as they focus on human problems of a region rather than maximum development of its resources. Two general comments are in order. Regional agencies do not in Latin America cover the whole of a country in a unified network. Instead in any one country there may be half a dozen agencies each with a different remit and method of working. This reflects the *ad hoc* way in which regional policy has evolved. Secondly, most policies to date have been within the diffusion-of-growth school. Alternatives have only been tried on a more or less experimental basis in limited areas.

Growth poles

Taking the relatively narrow policies first, those for industrial development mostly come under the rubric of growth poles or growth centres, while most agricultural ones have involved some colonization. Growth pole policies have enjoyed considerable popularity from the late 1960s to the present, though few have been successful, which is hardly surprising since they have been copied with little modification from their European forbears. In Europe, it has been possible to influence industrial location by a series of financial inducements and restrictions placing high taxes or physical limits on growth in the metropolitan area, while advancing credits and freedom from tax to industries moving out to selected cities, the new growth poles. In Latin America, these financial weapons have not controlled industrial location, because many companies are able to avoid paying tax and to disregard planning regulations, and more important factors are the proximity to government offices and administrators in the metropolis, as well as the far better skilled labour market, accessibility to major ports for imports and exports, and especially for foreign investors, the lower perceived risk in a central location (see Ch. 5).

In Argentina for example, regional industrial policy began in the 1920s with free port status (i.e. no import duties) for territories beyond the 42nd parallel of latitude, but this attracted no new industry apart from small meat and wool processing plants for Patagonian sheep. This did nothing to reduce the enormous concentration of industry at Buenos Aires city – in 1948 55 per cent of value added in manufacturing was at Buenos Aires – and further measures were tried in the 1965 plan, industrial parks at Cordoba and San Nicolas. This policy was extended by the 1970–75 plan which called for growth poles at various places in Argentina, which might have had some success at one level, that of expanding industry in the chosen sites, had their functions been better chosen and the appropriate environment for industry been created by provision of infrastructure.

In the event little was achieved because the second Peron period intervened from 1973, with a new orientation, but little could have been expected in the poles themselves because of the lack of investment by state or private firms, because of the poor choice (in terms of direct or induced employment) of the growth pole industries, and because the locations chosen were themselves not viable (Fig. 6.2). In the north-west, a single pole city could not be found and a group, Salta, Jujuy, Güemes and San Pedro, were to serve, towns which had few links between them. In the north centre, it was Corrientes and Resistencia, on either side of the Río Parana; in the north-east, Posadas was the pole with neighbouring towns, where development would be founded on hydro power and pulp and paper mills, not a large employer nor linked to other local industries. As for the three remaining poles, they were located in Patagonia, where no good case could be made for growth poles as population was scanty and not impoverished (Morris 1975). This was a case of putting place development above people development, and national security above either of these – Argentina still views Chilean claims to all or part of Patagonia with mistrust.

Elsewhere in the continent urban industrial policies which have been termed growth poles include Venezuela's Ciudad Guayana and Brazil's Brasília. Both of these illustrate the point made with reference to Argentina about location. Ciudad Guayana grew as a provider of electricity for the national grid, of steel and of

Fig. 6.2 Argentine growth pole cities 1970–75. Tucumán is shown, though not designated, as a growth pole, as it received special treatment following the partial collapse of its sugar milling industry

aluminium and paper for the nation, but was located far from the densely populated and poor northern states where peasants still engage in subsistence farming. Brasília was intended partially as the growth pole for the whole Centro-Oeste region, but it served far more the symbolic role of Brazil's *marcha para oeste*, the expansionary march to the west, and the turning of attention away from Europe and the ocean which reminded one of colonial links. As an agent to promote colonization or economic development it was by all accounts extravagantly expensive. Perhaps the best that can be said for the growth poles is that they were better than a simple industrial growth policy. In Mexico, the policy of tax relief which attracted industries to towns on the northern borders between 1965–72, was extended to all Mexico in 1972, with the result that industries and labour crowded even more into the capital.

Colonization

This has for long been a concern of Latin American countries, because many of them have densely peopled core areas, but peripheries which are demographic deserts. Whole states of northern and western Brazil have had population densities under two per square mile, which are only now being exceeded. Most of the colonization has been unplanned and undirected, the gradual unseen drift of peasants from overpopulated and eroded farmlands to new frontiers. But much recent colonization has been under the auspices of government in a direct or indirect manner. The best known and most important schemes have been around the Amazon Basin fringes, representing on the one hand, the down-hill, eastward push from the densely peopled Andean Highlands, in Bolivia, Peru, Ecuador, and Colombia; and on the other the westward push of Brazilian settlers, a continuation of a centuries-old expansion of settlement by the Brazilian Portuguese at the expense of the Spanish claims.

It is impossible to summarize the diversity of colonization schemes undertaken among these countries, nor is it easy to make a critique of them along the line of thought pursued so far, i.e. that regional development schemes should attempt to reduce inequalities in income or welfare between regions. Much colonization is inherently 'place development', as it focusses on the occupation of new territory, unpopulated or with a tiny, often Amerindian population. There is here little question of raising living standards of large rural masses to national standards. Instead we may comment on two sets of problems: the technical ones connected with such schemes, and the colonization programme as a solution of national agrarian problems.

In the last century agricultural colonies were formed out of some of the huge estancias of the Argentine Humid Pampas and in the long term led in one area, central Santa Fé, to a certain difference in agricultural structure, with higher intensities of farming and an early subdivision of the great cattle estancias; but elsewhere there was very little permanent change, as the colonies were held only by tenant farmers who became re-absorbed into the cattle-ranching economy when landowners began to see them as a threat. In this century, an early scheme was that for North Parana, where a London-based company set up a planned colonization over an area of 12,000 sq. km (4,632 sq. miles), from 1923 to the 1950s. It has given a higher intensity of land use and road and rail development, but its ultimate success as a break of the typical Latin American structural problem of *latifundio/minifundio* is uncertain; many of the farms, first set up as owner-operator units, have reverted to the typical tenant status.

More recent evidence comes from the tropical lowlands areas, an example of which is the area around Santo Domingo, Ecuador (Fig. 6.3), the object of colonization exercises since the 1960s (Lowder 1982; Wood 1972). It has attracted colonists fleeing the droughts of the southern border lands and the west coast (PREDESUR and CRM areas), farmers displaced by agrarian reforms, and plantations workers attracted by the 1960s banana boom, followed by oil palm and manila hemp plantations. In this area there have been both physical and human problems. Typically, the physical problems centre around the inherent poverty of the soils, which deteriorate under monoculture even when the crop is a tree crop which provides some protection from sun and rain. Human problems have included the occupation of part of the area by large farm companies which have efficient plantation operations in competition with small farmers.

Other problems, repeated so many times in the tropical lowlands, have been those of inadequate infrastructural provision. Roads, marketing systems, farmer credits, technical advice, schools, hospitals, rural service centres, have all been deficient or in some areas totally absent, the 'colonization' being limited to a division of lands into lots and their distribution to settlers.

In addition, the colonization areas have become differentiated so that now 32 per cent of the farms are under 20 ha (49 acres), probably a minimum size for commercial farming in the area, while 12 per cent of the land is held in units of over 500 ha (1,235 acres). As a result, though it is impossible to comment on development relative to the rest of the country, a sharp internal differentiation is now observable, between small districts within the region. From surveys done by local government and the Interamerican Development Bank, variation between districts is from over $800 annual per capita income, to under $200, effectively a range between successful commercial farming and near-pure subsistence.

Summarizing this section, one major set of problems relates to technical deficiencies plus the lack of careful administrative control from the time of the first colonies. Some of these difficulties are answered by the newer approach of Integrated Rural Development, to be commented on shortly. The other set of problems arises from

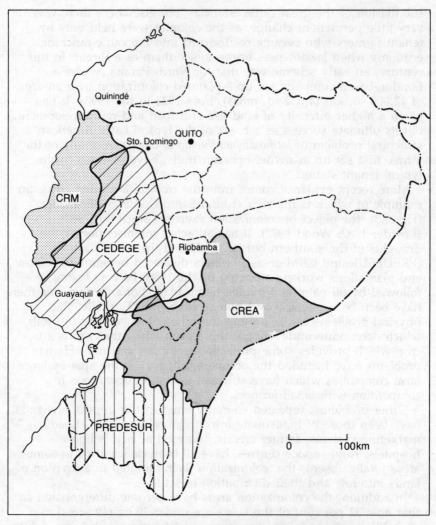

Fig. 6.3 Areas of jurisdiction of the Ecuadorean regional agencies. As elsewhere in Latin America, they form no national network but a disjointed and incomplete set with varying organizations and aims (based on Bromley 1977)

the failure of colonization to achieve the national aims set for it. As a means of solving the problem of burgeoning population on the old densely-peopled Sierra lands, by moving them out to the lowlands, it has been a conspicuous failure, for the number colonizing the lowlands has never been equal to the annual increase of Sierran population in Ecuador. In terms of equalizing incomes between major groups and areas, it has been noted that poverty is not excluded from the colonization areas, and it has often been reported that the problems of the Sierra, in all the Andean countries, are repeated at a generation's distance in the lowlands, when tenancy, landlessness, exploitation by merchants buying the products, re-emerge in the newly-settled lands.

Problem Regions

This title covers a somewhat variable set of policies which have grown up in different ways to meet the needs of different kinds of crisis regions. An agency that may be looked at in a little detail is SUDENE in Brazil. SUDENE emerged in 1959 in north-east Brazil from previous institutions set up to alleviate the effects of massive and recurrent droughts, especially in the *sertão* region, the semi-arid lands lying 300–500 km (200–300 miles) inland, a physical problem underlain by inadequacies of the economic and social system, such as insecurity of rural employment, small farms, lack of food crop production, and lack of alternatives to agriculture in the region.

In the north-east early policies were for famine solution – disastrous droughts in the 1870s were attacked by distribution of food in the cities, Fortaleza, Recife, Bahia, to which rural migrants had come. This 'Oxfam' approach was broadened a little by the creation of work projects to employ migrants and the newly unemployed. From early in this century, the policy became a 'hydraulic policy'; as the problem was perceived as drought, the drought should be subject to frontal attack, by building reservoirs on the rivers and by digging wells. There were many technical deficiencies in this policy, notably the failure to build dams of large enough size to mitigate the effect of the summer drought, to prevent their filling in with sediment, or to construct canals to the lands of small farmers where the water might be used. The grosser failures, however, were due to the lack of any attack on the human problems such as a semi-feudal agriculture, and an absence of provision for basic human needs in food and shelter.

From 1959, SUDENE acted as regional co-ordinator for development projects throughout the states of the north-east, from Bahia north to Maranhão. The emphasis changed again, from 'hydraulic' schemes to an investment approach – industries were to be set up in industrial parks, and to be encouraged to come to any part of the north-east, while agriculture was to become more productive and effective in meeting local demands, i.e. more subsistence crops were to be grown in place of the dominant cattle-breeding, cotton, sisal, and other commercial crops of the *sertão*. Agency activities would be accompanied, it was hoped, by a real agrarian reform that would put more land in the hands of small farmers who needed it, and out of the presiding system of latifundios and plantations. In industry, the only advances were made in the three great cities, Fortaleza, Bahia, and Recife, where incoming industry was concentrated. In agriculture, little was accomplished because there was no agrarian reform.

157

In the event, SUDENE's policies swung from an initial concentration on industrial growth to a later emphasis on agriculture, and the hydraulic policy was never totally dropped; the latest phase has been the downgrading of SUDENE's activities and its subjection to considerable central control under the military government. The leading element of regional policy became colonization of the Amazon and the roads programme. SUDENE may be regarded as an agency with a good aim – in terms of the stance taken earlier since it was directed at a heavy concentration of population living in the greatest poverty. The north-east states were the poorest in the whole of Brazil, and population density much higher than in the other peripheral areas, the north and centre-west. A 'misery index', an index of relative need composed of population density multiplied by the level of income in the region, would indicate this region as the focal one for attention in Brazil. Given the breadth of its approach, many of the technical difficulties it faced could no doubt have been overcome, by redirecting industry to rural areas and by improving the quality of its attack on drought. More difficult to attack would have been the lack of agrarian reform of an effective nature, and in recent years, the lack of central government commitment to the agency.

River basin policies

These have also generally been set up as broadly based approaches, following the Tennessee Valley model, involving agencies in power production, flood control, erosion control by afforestation and farming improvements in the basin. Like colonization, they may be regarded as 'place development' because of the need to focus on the physical unit. Mexico is the best Latin American example of such a policy, having created several agencies covering river basin development over most of the south of the country (Fig. 6.1). As in the USA, the river basin approach was a way for Federal government to involve itself in the regions while avoiding power struggles with the individual states, since the river basins have boundaries which differ from those of the States, and generally include more than one state.

These agencies came into being partly as a new initiative on the rural problem which had not been solved by the agrarian reform of 1917, and revivification of communal farm ownership through the *ejido*. Now, instead of agrarian reform, the river basin approach would bring colonization of the tropical lands, along with new industries using hydro power. The first attempt, from 1947, was with the Papaloapan river in the south; flood control of this powerful river was seen as of first importance, to free the lower valley for agricultural uses. Power production would be used in timber working and paper manufacture, and roads and settlements would be created in the lands made usable by flood control. These aims were only partly fulfilled, and new problems were created.

Production – of sugar, of cattle, of timber and paper, of power – was certainly increased, but small farmers were not helped and the irrigation projects failed. The cattle came from 770 ranches in the lower valley, from seasonally flooded natural grasslands which were simply improved by flood control. Sugar production came from commercial farms and estate lands to feed a giant mill and supply one third of national sugar needs. Paper production was also based on large scale production and for outside markets. Somewhat similar critiques could be applied to the other regional development schemes, the Tepalcatepec Basin, begun in 1947, and from 1960 incorporated into the Balsas, the Lerma-Santiago-Chapala (1950), the Fuerte (1951), in the Pacific north-west, and the Grijalva-Usumacinta scheme just east of the Papaloapan, where work began in 1953. Most of the states involved would not seem to have benefited very much from the investments in terms of raising regional incomes. The states of the Papaloapan basin, Oaxaca, Puebla, and Veracruz, had GDPs per capita in 1960 which were 35.4, 55.3 and 74.6 per cent of national average respectively, and in 1980 these figures had improved only marginally to 37.4, 60.8 and 78.6 per cent of national levels, despite regional development, and in the case of Veracruz, a continued growth of the oil industry.

At one extreme, some critics of these schemes imply that they were set up largely to engross the coffers of major firms and wealthy investors and merchants, who benefited from the profits on projects such as building dams and roads, or from the new production of commercial crops for export out of the region. What seems more likely is that there was a good deal of personalism in the choice of projects undertaken and their location. The Papaloapan scheme was particularly associated with President Miguel Alemán, the main dam named after him as well as the main new settlement in the area; Alemán had been governor of Veracruz and was a native of the area, and the whole project was thus a kind of monument to him. Many private firms and institutions have been the major beneficiaries, but not so much by state design as through the lack of regulations controlling their activities, the lack of an effective agrarian reform, or of an effective progressive taxation which would reduce inequalities. In other words, through the combination of structural difficulties outside the control of the projects' directors.

At the other extreme, some would blame failure largely on technical faults. Intensive irrigation agriculture as the central objective, failed because of lack of knowledge about relevant crop types and insufficient attention to detail points like the supply of credit to small farmers, the failure to link agriculture to industry, and insufficient commitment to the project over a long period of gestation and development. It is certainly true that policy moved away from river basins to sectoral programmes of rural development, industrial development through financial attractions, which led to strong spatial biases (the concentration of industrial developments in

Mexico City, and of agriculture in the northern irrigated zones, which were already the most productive). A more balanced view is that both technical and institutional weaknesses have been present in Mexican policies to date.

Integrated development

Integrated rural or rural-urban development, has emerged in the 1970s using some of the critique of previous efforts, especially their urban bias, and including some of the ideas which were incorporated into 'development from below'. An example from Ecuador will illustrate the kind of project.

Comprehensive rural development was attempted in Ecuador, Peru and Bolivia under the United Nations Andean Mission. In Ecuador the Mission actively promoted Community Development from 1956 to about 1970. The theme of the programme was integration of isolated rural communities, and the approach was to be multi-faceted, involving agriculture, artisan crafts, transport, housing and education, as well as community organization, In Ecuador, 231 communities were affected, but this only meant some 6 per cent of the Sierra population, and the funds applied were tiny and dispersed over a huge variety of projects. It is difficult to assess the effects of such a programme because of this very diversity of actions, but it must be seen as less successful than hoped. Apart from lack of funds and commitment by government, the failures may be attributed to some 'structural' features. The local mayor and council, often the community's most infamous exploiters, were placed in charge of local projects; and agrarian reform had not been carried out in most communities affected. What seems best revealed is the need for any 'grass-roots' or 'community' development scheme to have available large numbers of skilled workers, whether local or outsiders, able and willing to dedicate much time to working in isolated rural environments. The number of such workers was always inadequate in the Ecuadorean Sierra.

After 1970, Ecuadorean regional policies moved away from the small-scale, integrated ideal, towards much more selective investments and massive schemes. CRM in Manabi, PREDESUR in the southern Sierra, CEDEGE in the Guayas valley, all centred their programmes on massive dam construction, hydroelectric and irrigation schemes, influenced no doubt by the pressure of the international aid agencies to have precise economic cost/benefit analyses done and technical plans which would identify the application of all loan funds given. This left only CREA growing out of its early interests in all rural sectors, to apply a broader approach.

In the late 1970s, interest was re-awakened in this broader approach, culminating in the establishment of a national secretariat (SEDRI) for Integrated Rural Development, in late 1980. One of its projects was Quinindé, a newly emerging problem area in the north-west Pacific coast lowlands. A small negro ex-slave and Indian population lived along the rivers, and there were a few cattle ranches, until the 1960s, when massive colonization began, by migrants from the economically depressed banana region of Santo Domingo, by migrants from the Sierra and by two large firms who bought land, one to produce plywood, the other for the production of African oil palm.

The Quinindé Project was not 'development from below', in any strict sense, since its initiative came largely from governmental agencies, at local, provincial and national levels. It attempted however, to work towards local decision making by starting with seminars and workshops in the area in 1977 to find out what were the perceived priorities for action. This gave rise to a great diversity of possible projects, only a few of which could be faced immediately: marketing of local produce via a co-operative was one, in order to circumvent the exploitative middlemen who offered low prices to producers but sold high; another was overall aid to the Cayapa Indian tribal group, and a third technical aid and credit for small farmers, focussing on the co-operatives rather than on individuals.

This scarcely constitutes an 'integrated' development programme in terms of products and their linkages, though more complete programmes will be built up. It does, however, pay attention to some of the ideas now current. First, it focusses on the relief of poverty, rather than on maximum overall growth of the economy. It concentrates on rural areas rather than any towns, reversing normal tendencies. It also seeks to provide for basic needs inasmuch as these were identified by the locals. The process of regional planning was also substantially modified. Instead of the usual regional plan, which focusses on the complete plan document at first requisite, then seeks funding, then moves on to (usually very partial) completion, the plan emerged in a process of planning and execution, not a document, so that modifications and additions were made continuously in response to local needs. This was seen as an important matter by the planners to allow proper local participation. It also obviously brings various problems in organization and justification before authorities, and its long-term success is by no means yet assured.

We might note, however, that some of the elements of development from below will be hard to apply in Ecuador, or indeed anywhere else in Latin America. In Quinindé, there is a certain flexibility because settlement in this area is itself mostly recent. In the Andes, however, developmental planning is constrained by the existing social and cultural structures. Towns, even quite small ones, are centres of power and privileged groups. Some local statistics may illustrate this matter. Guamote, South of Riobamba in the Ecuadorian Sierra (Fig. 6.3), is a town of 2,500 with a rural hinterland of poor farmlands, 50 per cent of it windswept *paramos* of rough grassland lying at over 3,800 m (12,467 ft) (Iturralde 1980). In Guamote district there is a racial and cultural division between town and country: one third of the Indian population speaks only Quechua. The urban population is 80 per cent *mestizo*, 20 per cent Indian, while rural population is 98 per cent Indian.

These divisions match a division in employment and in income levels. Only 1.9 per cent of agriculturalists are *mestizos*, the remainder are Indian. Conversely, 79 per cent secondary and tertiary employment is dominated by *mestizos*. It is these kinds of structural problems that will prove difficult to solve by local 'development from below' projects, which must depend on local leadership, and which would thus be absorbed by and twisted to the interests of the leading *mestizo* group.

Government aims for regional policy

Superficially, regional policies would seem to have been established everywhere for the most humanitarian reasons; the differences between say, growth poles, colonization and river basin development schemes, would thus be no more than differences in the means to a common end, the amelioration of the local human condition.

In fact, it is possible to discern various other purposes to regional policy: in most cases, these policies may be seen as a kind of legitimation, in other words, policies which present the government as benevolently inclined towards regions, when in fact it has little sympathy for them. Other more specific purposes are:

The monument policy
An example has already been quoted, that of the Papaloapan basin, associated with Miguel Alemán. Another Mexican example would be the Tepalcatepec Commission, headed by Lázaro Cardenas, former president of Mexico and a native of Michoacán, the state most affected by the agency. Other examples of this kind of policy are found in Brazil, most notably in the creation of Brasília in 1956–60, during the presidency of Kubitschek and very much a visual monument to his period.

National development through regional means
This driving force is present in most of the regional policies. Mexico sought to increase national raw materials production after the war in a period of development based on raw materials export and manufactured imports – *desarrollo hacia afuera*. Cotton, sugar and wheat production were substantially aided, though the regions did not greatly benefit. In Brazil, official plan documents give as a major aim of regional development, the production of food-stuffs to feed the ever-growing urban masses in the south-east. Government hopes were not to be realized in this matter, but the hope would appear to have been genuine.

National security and integration
This requires a little more comment as it is less obvious and perhaps less logical for non-Latin Americans. This formula has been a

watchword in the Brazilian National Plans during the military's rule (1964–80), and has been employed in the equally military governments of Argentina (1967–83), Peru (1968–80), Bolivia since the 1960s, and Chile since 1973. One focus is on the need for national economic self-sufficiency; thus regional development tends to given national independence of outside supplies. Thus development of the Concepción region of southern Chile relates to Chilean search for autarchy, in steel production, as does Peru's steel at Chimbote and Venezuela's at Ciudad Guayana.

National security also implies the need to occupy under-populated frontier regions in order to secure the national claim to those regions, often an uncertain claim for historical reasons. Argentina's Patagonian growth poles, Peru's or Bolivia's eastern colonization schemes, may be seen in a new light.

In the case of the Brazilian push to occupy the western regions, through a variety of regional plans, the logic may be more aggressive than simply the securing of presently agreed national territory. Brazilians have expanded their territorial control over the centuries using the principle of *de facto* possession and settlement as the basis of legal ownership, so that large slices of territory have been taken from all their neighbours. To maintain firm control over these areas and be able to resist any counter-move from neighbours, it was necessary to occupy the frontier, and to have good access to it, via the system of highways leading out to and also paralleling the frontier. The amount of attention given to this thrust into the heart of the continent, makes one suspect that defence of the frontier is not all. Brazil's neighbours have weak access to their border zones, and hardly represent a threat to their giant neighbour of 120 million inhabitants. Does Brazil wish to claim more land?

Such an idea might seem far-fetched but for the fact that Brazilian political thought since at least 1964, has been dominated by geopolitics, and a brand of geopolitics which follows in the German pre-1939 style. According to Brazilian politicians, the axial area of South America is in Bolivia where the north-south line of the Andes is crossed by the east-west interfluve between River Plate and Amazon tributaries. A power controlling this interior cross-roads can dominate both Pacific and Atlantic coast, and both northern and southern river basins; and yet there is a power vacuum in the region at the present. There are also allusions to the continental or maritime vocations of states. Brazil is seen as a basically continental state, though distorted by the colonial experience of linkage outwards to Europe. Thus she should now push boldly to the interior, and become the central state in South America.

There are even references back to Ratzel's nineteenth-century conception of the state as an organism; in this, the state acts as a live body, which is either in growth or decline, never static. If it is not growing, it is already moribund; Brazil, it is the dangerous conclusion, must grow territorially to confirm its strength and vitality. All of these ideas have penetrated through the *Escola*

Superior de Guerra, the War College, where both high-ranking officers and leading civilians in government have been trained. This prestigious institution has guided the overall philosophy of development, away from the SUDENE kind of thinking, in any case undesirable because it was too autonomous from the centre, towards the ideas of northern and western colonization.

The National Plan documents themselves include the following reasons for their regional plan elements:

1. to offset and relieve population pressure in coastal regions;
2. to rectify imbalances of regional development;
3. to provide more primary resources – minerals, power, food to the nation;
4. to promote colonization on the frontier;
5. to help Brazil towards its 'manifest destiny' in the interior.

There is no reason to doubt that these are real factors taken into account by government thinking. What may be said is that they do not admit clearly (only very indirectly in point 5) the geostrategy employed by Brazil and its interest in political and military dominance within the continent. Nor can there ever have been much credence placed in the possibility of locating many settlers out of the north-east in the interior, and thus relieving permanently the problems of poverty and overpopulation in the north-east.

Regionalism

Any reading of regional development theories suggests that it is dangerous to apply rich-country theory directly to poor countries. There may, however, be some relevance in the fact that in several European countries the regional issue has taken a different turn over recent time. In France, Britain, Spain and Yugoslavia, deep-rooted regionalisms have resurfaced in demands for devolution of power. When this happens, the socio-economic plans for regions by central government take second place or are superseded by the transfer of powers to the new regional authorities. Attraction is then focussed on how much and what kind of powers are involved in the transfer, which will in turn determine policies.

Regionalism, understood here as a sense of distinctive consciousness among a community smaller than the national one, has not been seen as important for Latin America. Regionalism may be cultural, economic, or politically based; cultural regionalism, linked to differences of religion, language or race, is certainly to be expected among such groups as the Andean Indians, but it has been suppressed since colonial times and now presents no threat to central government in the form of political movements. Historically the more important kinds of regionalism have been economic, with a partial political expression.

Regionalism may yet be important again in the context of the newer forms of development. If projects and programmes are to come from within a community or region, if emphasis is to be laid on local participation in selection, organization and fulfilling of plans, then an existing community is a good basis for success. In addition, if partial regional closure is desirable to allow the internal development process to go ahead, then some level of politico-administrative autonomy is perhaps a means for achieving closure; it is indeed difficult to imagine economic closure without such autonomies. A review of the most prominent regionalist movements of the past will show, however, serious difficulties in their use.

In Argentina, the interior provinces formed a federation from 1820–60, with different interests from those of the Humid Pampas provinces, especially Buenos Aires. Interior interests were in craft industries and food production for national markets only, thus demanding protection from the exterior, whereas Pampas interests were in free trade, stimulating the import of manufactured goods and the export of hides and other cattle products to Europe. These were interests mostly of the ruling class, merchants and industrialists and landowners, and the differences were conclusively settled in the centralized state of the 1860s. There remain cultural and economic differences, but not on a scale which generates demand for autonomy.

Similar separation of economic interest was behind the regionalist movements in Pernambuco, Rio Grande do Sul, and São Paulo in nineteenth-century Brazil. They were again the distinctive interests of small groups of influential merchants in states concentrating on individual export crops or products, cattle, sugar, or coffee. If such regionalisms are an unsatisfactory basis for modern development efforts, are there other spatial groupings which might serve? Communities do exist on a very small scale, in some areas, notably among the Indian or *mestizo* communities of the Andes and Central America. They have been incorporated in various ways and with limited success into previous developmental efforts – in Mexico the *ejido*, in the Andean countries the community development schemes of the UN Andean Mission, in Peru various elements of the co-operative system set up after the 1969 agrarian reform.

There are severe limitations on the role of these communities; they could not be used as the basis of industrial development except at the most basic level, they could not readily integrate large regions with urban centres, and in many areas the community structure has broken down beyond repair under the influence of new organizations and the market economy. In areas of little modern development or urbanization, however, they offer a social structure which is probably better than that of the modern co-operatives which are usually short-lived because of the lack of cohesion among members.

Other than these selected areas, there is an obvious need for something to replace the failed regional policies. This might be, not another regional policy, but an improvement of sectoral policies, which are currently too weak or have no redistributive effect. It is necessary for example, to balance policies protecting local manufacturing with tariffs (a policy which encourages growth in the largest cities), with policies encouraging raw material production, especially for regional or national markets, as such policies will commonly help remote rural areas. Tax policies which move from the present bias on indirect taxation towards direct (income) taxes, which tax effectively and progressively, are another measure with considerable regional impact in countries of strong inter-regional income differences. In many countries large companies, e.g. in the oil industry, are owned by the state or state agencies, and a large element of regional development is determined by their policies. Under these kinds of conditions, regional policies could be a useful addition, to cater for regional peculiarities once an equitable general environment has been provided.

Further reading

Regional development generally and in Latin America

Cardoso F H and **Faletto E** (1979) *Dependency and Development in Latin America*. University of California Press, Berkeley and London. (This is one of very few texts by Latin American dependency writers translated into English, and gives a balanced account.)

Gilbert A (1976) *Development Planning and Spatial Structure*. Wiley, Chichester and New York. (This set of essays shows various kinds of inter-relationships between planning efforts and geographical space in the Third World, using Latin American and other examples.)

Gilbert A (1978) 'The state and regional income disparity in Latin America', *Bulletin of the Society for Latin American Studies* **29**, 5–30.

Gonzalez-Casanova F (1969) 'Internal colonialism and national development', in **Horowitz I L, de Castro J** and **Gerassi J** (eds.), *Latin American Radicalism: a documentary report on left and nationalist movements*, pp. 118–139. Vintage Books, New York.

Lo Fu-chen and **Salih K** (1981) 'Growth poles, agropolitan development and polarization reversal: the debate and search for alternatives', in **Stöhr W B** and **Taylor D R F** *Development from Above or Below*, pp 123–152. Wiley, Chichester and New York.

Morris A S (1975) 'Regional disparities and policy in modern Argentina,' Occasional Paper No. 16, Institute of Latin American Studies, University of Glasgow, Glasgow.

Morris A S (1981) *Latin America: Economic Development and Regional*

Differentiation. Hutchinson, London. (An attempt at viewing economic development alongside other, non-market processes over time, based on Latin America, with recommendations for spatial planning.)

Stöhr W B and **Taylor D R F** (1981) *Development from Above or Below*. Wiley, Chichester and New York. (The best-known 'text' on development from below, though actually a series of contributions from a wide variety of writers; uneven but very stimulating.)

Mexico

Barkin D (1975) 'Regional development and interregional equity: a Mexican case study', *Latin American Urban Research* **5**, Ch. 12, 277–299. (Fills out the Mexican experience. Highly critical of regional development achievements.)

Barkin D and **King T** (1970) *Regional Economic Development: the River Basin Approach in Mexico*. Cambridge University Press, Cambridge. (Gives a detailed account of the river basin approach in Mexico and some critique of its achievements.)

Friedmann J, Gardels N and **Pennink A** (1980) 'The politics of space: five centuries of regional development in Mexico', *International Journal of Urban and Regional Research* **4**, 319–349. (Tries to bring together varied elements from Mexican regional history to show a common thread of development efforts.)

Ginneken W van (1980) *Socio-economic Groups and Income Distribution in Mexico*. Croom Helm, London. (The Gini index used is calculated as: $\frac{1}{2}\Sigma(x - y)$, where x is share of GDP, y is share of economically active population.)

Scott I (1982) *Urban and Spatial Development in Mexico*. Johns Hopkins University Press (for the World Bank). Baltimore and London. (A good assemblage of regional and urban data not critical of the basic aims and achievements of Mexican planning. Attempts a spatial plan for the future of Mexico.)

Ecuador

Bromley R J (1977) *Development and Planning in Ecuador*. Latin American Publications Fund, London. (This describes the main agencies for regional development in Ecuador and the context of Ecuadorian developments.)

Iturralde D A (1980) *Guamote: Campesinos y Comunas*. Instituto Otavaleño de Antropologia, Otavalo, Ecuador.

Lowder S (1982) 'La colonizacíon como estrategía para el desarrollo: el caso del cantón de Santo Domingo de los Colorados', in **Ryder**

R and Robelly N, *Geografía y Desarrollo*, pp. 127–234. CEPEIGE, Quito.

Morris A S (1971) 'Spatial and sectoral bias in regional development: Ecuador', *Tijdschrift voor Economische en Sociale Geografie*, **72** 279–287. (Criticizes the kind of results being obtained by the agencies, and contrasts the large-scale approach commonly used to efforts close to 'development from below'.)

Wood H (1972) 'Spontaneous agricultural colonization in Ecuador', *Annals of the Association of American Geographers* **62** (4), 599–617.

Major statistical source

Wilkie J and **Haber S** (eds) (1983) *Statistical Abstract of Latin America* **22**, UCLA, Latin American Centre Publications, Los Angeles.

CHAPTER 7

Urban growth, employment and housing

Alan Gilbert

Urban growth in Latin America

Rapid urban growth is hardly a new phenomenon in Latin America. In Argentina, Chile, Uruguay and southern Brazil, several major cities had emerged by the end of the nineteenth century (Table 7.1). Successful export production in these areas had led to the concentration of transport and commercial facilities in cities such as Buenos Aires, Montevideo, Rio de Janeiro and São Paulo. The pace of expansion was fuelled in each case by major flows of immigrants from Europe especially from Spain, Portugal and Italy (see Ch. 9). By 1920, Buenos Aires had 1,600,000 inhabitants and Rio de Janeiro 1,200,000.

Elsewhere in Latin America urban growth had begun by 1900 but was occurring much more slowly. Indeed, in the smaller Central American and Andean countries, where industrialization did not really begin until the Second World War, urbanization only started to accelerate during the 1930s. This rise was associated less with the pace of economic growth than with a general fall in the death rate. From the 1930s, the populations of most Latin American countries were increasing rapidly. At first, most of the natural increase was concentrated in the rural areas and was largely absorbed there, but, when population growth rates rose to over 3 per cent per annum, industrial development accelerated, opportunities for employment in the cities increased, and as land redistribution in favour of the poor failed to materialize, rural to urban migration got under way. By 1970, two out of five Latin Americans lived in cities with more than 20,000 people; in Argentina, Chile, Uruguay and Venezuela nearly two-thirds of the population fell into this category.

The migrants were attracted to the cities by the availability of better paid work and encouraged to move by the increasing difficulty of making a living from small-scale farming. The opportunities available in the cities and the level of rural poverty obviously varied from region to region, but, even where there were major differences in rural and urban living standards, many people

Table 7.1 Urban population in towns of 20,000 inhabitants and over, 1900–70 (%)

Country	1900	1920	1940	1950	1960	1970
Argentina	24.9	37.0	41.0	49.9	59.0	66.3
Bolivia	6.6	9.4[a]	18.7[b]	19.4	22.9	27.2
Brazil	8.7	13.0	16.0	20.3	28.1	39.5
Chile	19.9	27.6	37.0	42.6	50.6	60.6
Colombia	6.2	8.9[c]	14.0	23.0	36.6	46.2
Costa Rica	8.5	12.2	15.4	17.7	24.4	27.0
Cuba	25.0	23.2	32.2	36.1	38.9*	43.4
Dominican Rep.	3.6	4.0	7.0[d]	11.2	18.7	30.2
Ecuador	9.3	13.6	17.0[e]	17.8	27.9	35.3
El Salvador	6.0*		12.4	13.0	17.7	20.5
Guatemala		9.2	9.8	11.2	15.5	16.0
Honduras		6.3	6.1	6.8	11.5	20.2
Mexico	9.2	12.6	18.0	23.6	28.8	35.1
Nicaragua		10.9	21.8	15.2	23.0	31.0
Panama	6.8[f]	21.9	25.0	22.4	23.1	39.4
Paraguay	13.0	18.0	15.7	15.3	15.9	21.5
Peru	6.0[f]	5.0	12.9	18.1*	18.5	40.3
Uruguay		27.8[g]	36.8[h]	53.1*	61.4	64.7
Venezuela	8.5	11.7	28.5	31.0	47.0	59.4
Latin America**	10.9	15.5	20.5	28.2	33.0	41.5

* Estimate
** Weighted by population in each country

a	1923	e	1938
b	1942	f	1903
c	1918	g	1922
d	1935	h	1937

Source: Wilkie and Perkal (1983: 140)

simply stayed where they were. This was especially true of the Indian-speaking populations of Ecuador and Peru, groups which would suffer most economic and social discrimination in the urban environment. Certainly, the population of the rural areas of northern South America continued to grow in absolute terms until the 1970s.

The migration process was also highly selective within the rural areas. Young people who wished to further their education, new entrants into the job market, persons with skills such as bricklayers and drivers, were more likely to move than older and unskilled workers. In most areas more women moved to the urban areas than men, largely because of the opportunities for unskilled labour available in domestic service and retail activities in the cities. Indeed, the differences in the numbers of men and women migrating to Latin American cities was often quite marked. In Bogotá, in 1973 there were eighty-seven men to every hundred women and among the main migrant age groups, 15–45 years, there were only eighty-two men (Colombia, DANE 1977).

Practically all urban centres gained from rural out-migration and since 1940 most cities have grown faster than the rate of natural increase. Some small cities have experienced dramatic expansions in their population. In Peru, Chimbote grew from a town of 5,000 people in 1940 to one of 173,000 in 1972, a result of a boom in fish-meal production and the establishment of a new steel works. In Venezuela, the population of the new industrial city of Ciudad Guayana increased from 3,000 people in 1950 to 320,000 in 1981. In Colombia, agricultural frontier towns like Montería, Valledupar, Villavicencio quadrupled their populations in the twenty years after 1951 and along the Mexican–US border towns like Tijuana, Mexicali and Ciudad Júarez saw their populations increase four or five times between 1950 and 1970. In terms of sheer numbers, however, the most spectacular increases have been in the largest cities. Today, Latin America contains several of the world's largest cities; in 1980 Mexico City had approximately 13,600,000 inhabitants, São Paulo 12,300,000, Rio de Janeiro 9,600,000 and Buenos Aires 10,200,000. Even previously small Andean cities have become major metropoli, Caracas has 3,200,000 inhabitants, Bogotá 4,500,000 and Medellín 1,700,000 (Fox 1975). What is perhaps more startling is the pace of change. Between 1970 and 1980 Mexico City added some 5 million people to its population, São Paulo 4.5 million.

Many among the new city dwellers were migrants, but, as the years have passed, much of the pace of urban expansion has been maintained by births in the cities. Since so many of the migrants have been young they have raised families in their new homes; no longer is rapid urban growth due only to migration. Indeed, the day has been reached in many Latin American cities when migrants have ceased to make up the bulk of the population. This process of change can be seen by the experience of Mexico City. During the 1940s more than two-thirds of its growth was due to migration, but, by the 1950s, the proportion had fallen to two-fifths, a figure that was maintained during the sixties (Unikel 1976). Whatever the problems of that city most of the population have been born and bred there.

Not only has population growth in the major cities of Latin America been spectacularly fast, it has also underlined the degree of urban concentration within each country. The share of the national population living in the largest cities has continued to grow. By 1970, two out of five Argentinians lived in Buenos Aires, half of all Uruguayans in Montevideo, one in four Chileans in Santiago and one-quarter of all Costa Ricans in San José. These cities have long dominated the economic and political life of their nations and the continued movement of people to these urban areas has confirmed their domination. Any number of economic variables can be used to show how the large cities of Latin America dominate their nations. In per capita terms there is more industry, more financial activity, more investment and higher incomes than in other cities. In 1972 Mexico City, with 22 per cent of Mexico's population in 1970, consumed 55 per cent of the country's finished industrial products and contributed 36 per cent to the gross domestic product (Unikel 1976).

Given the nature of economic development in Latin America it is not surprising that these major cities have increased the level of primacy within their national economies. Many are major ports (Rio de Janeiro, Lima-Callao and Guayaquil) and some also dominate the most productive agricultural regions of their countries (São Paulo, Santiago, Buenos Aires and Montevideo). Transport routes have spread outwards from these urban complexes giving them a major competitive advantage. In addition, since most higher income groups are concentrated in the major cities, they dominate the market for cars, refrigerators and clothing. Many of these cities are the seats of national government and have benefited from the unprecedented expansion of the public sector during the post-1945 period. Finally, as governments have increasingly adopted an interventionist role in the development process, the numbers of workers in the civil service, in public utilities and in public transportation have increased. As some vestige of a welfare state emerged, more teachers, health workers, and public administrators were employed. Not only did the location of government offer companies a market but it also eased the process of petitioning for import licences and price rises. The result was that commerce and industry became more and more concentrated in the major city regions.

Many governments have expressed concern about the pattern of urban concentration. For a start, the simple arithmetic of population growth is a spectre to the governments of the continent; forecasts of Mexico City's population in the year 2000 regularly conjure with the figure of 30 million. Given the current problems of traffic congestion, air pollution, lack of water and rising land values this is an understandable concern. Already it is extraordinarily difficult to move about cities such as Caracas and the levels of air pollution in Mexico City and São Paulo would not be tolerated in most cities of the developed world. In addition to these disadvantages, concern is frequently expressed about the difficulties that the expansion of these major cities puts on the rest of their nations. Provincial politicians regularly complain that their regions are neglected by the concentration of investment in the major cities. They point out that the largest cities are growing most rapidly; while particular provincial cities and regions have grown spectacularly, most have lower growth rates than the major cities. Living conditions are also inferior in most smaller cities because levels of investment in public utilities are much lower. Similarly, there are constant complaints that the major cities are extracting wealth from the rest of the nation: provincial taxes exceed central government spending in the regions;

skilled manpower leaves for the large cities; centralism drains local autonomy and provincial dynamism.

However, if there are innumerable complaints about metropolitan growth, many advantages may also derive from this pattern of urban agglomeration. Indeed, given the commitment of most Latin American governments to a combination of modernization and industrialization within a capitalist model of development, metropolitan expansion is almost inevitable. For many in the metropolitan centres there are few problems. When governments have attempted to decentralize manufacturing, for example, industrialists have rarely been sympathetic. Even plans to move government workers away from the major city have been problematic: the move to Brasília was hardly popular among Brazilian civil servants and, in Colombia, top administrators of one state corporation resigned rather than move to a provincial city. In Venezuela, the director of the Guayana Corporation still has his office in Caracas; it is the deputy director who works in Ciudad Guayana.

This opposition to relocation is due to the fact that living conditions are inferior in the provinces. Indeed, living conditions are normally superior in the major cities. Power supplies, water systems, housing and services are normally better than those available in provincial cities. In Colombia, for example, the proportions of households lacking piped water in 1973 varied from about 5 per cent in Bogotá, Medellín and Cali to around 60 per cent in cities with less than 200,000 people (Colombia, DANE 1977). Average incomes are higher even when allowance is made for higher costs of metropolitan housing. It is probable that housing conditions for the poor are more cramped and fewer families can afford their own home, but the catastrophic living conditions that are often described in the major cities are less true of life for most people in the large than in the smaller cities.

Of course, living conditions in the largest cities should be improved but the critical question is how best to achieve this. One answer is to decentralize activity from the major cities and to discourage further concentration; another is to plan the process of metropolitan expansion more effectively; a third is to modify the model of development in such a way that lower income groups throughout the country share in the benefits of economic growth.

It is less than clear whether urban deconcentration is the answer in Latin America. Numerous attempts to deconcentrate industrial activity over the past thirty years have produced few notable successes. Many of these attempts have been expensive and they have created few new jobs in the recipient regions. Even in north-east Brazil, the Venezuelan Guayana and the Mexican state of Guerrero, large new industrial complexes have brought all too few benefits for the poor (Gilbert and Goodman 1976; Travieso 1972; CEU 1977). Sometimes the results of deconcentration have been almost laughable, for example, the incentives given to the Chilean car industry to develop in the far north of the country (see Ch. 6). More often than not the policy has just not worked.

Of course, industrial deconcentration may be inevitable in places. In Caracas and São Paulo government controls on pollution levels are encouraging movement away from the main city, although most of the relocated industry is establishing itself within the metropolitan region rather than in peripheral parts of the country. Urban deconcentration as a means to improve living conditions in the urban areas is not the same as bringing benefits to the poor regions of the country. For the poor regions to benefit there also needs to be a programme of rural development which embraces land reform and local participation in decision-making. As experience in the Brazilian north-east shows merely enticing capital-intensive industry to poor regions does not help the rural or even the urban poor (Gilbert and Goodman 1976).

Whether or not deconcentration of economic activity is necessary, better planning is required in the larger cities. We say more about planning below but it is clear that present efforts at metropolitan planning are sadly deficient. Planning regulations are ignored both by the private sector and by public agencies. The essential problem is that good planning is based on compromise and compromise is not easily reached in Latin America. In Latin American cities there always seem to be good reasons why absurdities should be allowed to continue. In Bogotá, private transport companies refuse to introduce bus stops because they maintain it will cut their takings; the effect on traffic flows and accident rates as buses swerve across lanes to the kerbside to pick up additional passengers can be imagined. Urban land-use regulations are ignored in the poorer parts of most cities because if they were enforced, most housing would have to be demolished. Clearly, improved planning of transportation, housing and servicing is even more essential as urban populations increase. Gradually, there are signs that this is emerging. Many cities which have requested loans from the major international lending institutions have been forced to modify their planning processes. Most of the major electricity and water improvement projects that have occurred in Latin America's larger cities have been financed from abroad and loans have often been subject to strict regulations about pricing, organization and efficiency within the public utilities (World Bank 1983; Gilbert and Ward 1985). Whether, of course, pressure of this kind is a wholly positive development is a fascinating area for debate but change is certainly occurring to rationalize the provision of these services.

The major solutions to metropolitan expansion, however, lie outside the realm of either regional or urban planning. They lie in changes to the model of development that has been adopted in most Latin American countries; many of the urban problems are a direct consequence of national policy decisions. Acute traffic congestion, for example, is a corollary of the decision to maximize domestic car production by encouraging home consumption. Poor housing

Plate 10   Central Caracas, where metropolitan expressways and towering office blocks cover the valley floor while low-income settlements occupy the lower slopes of the hills in the distance. (Photo: A. Gilbert)

conditions are part of the price that must be paid for the refusal to redistribute income and to claw back rapid rises in the price of urban land. The very process of metropolitan growth is an outcome of the failure to redistribute agricultural land to the peasantry, the decision to favour urban activities in investment decisions, and the preferences for building public housing projects in urban rather than in rural areas. Urban consumers receive cheap foodstuffs at the expense of small-scale rural producers; migration to the city is one result. Clearly a redefinition of the priorities of the development model would not halt metropolitan growth. Nevertheless, it would slow the current pace of expansion and, more importantly, would resolve many of the problems facing the poor both in the urban and the rural areas. Whether or not this is a likely change is beyond the scope of this chapter but it is important to realize that the present distribution of the population and the conditions in which they live cannot be treated only through the tools available to urban and regional planners.

Employment and the distribution of income

The most critical problem in Latin American cities is the lack of sufficient well-paid jobs. Were more such jobs available, the problems of housing, nutrition and health would be much reduced.

175

Of course, many in the urban areas have regular and well-remunerated work. In addition to adequate salaries they also receive benefits such as health insurance, social security, retirement schemes, and access to sports facilities. Most of these people work in the so-called 'formal sector' which includes the larger industries and offices, the bigger shops and commercial enterprises.

While there is no watertight definition of this sector, commonly used criteria include the following: corporate ownership, large scale of operation, capital-intensive technology, government-protected markets, formally acquired skills, dependence on foreign techniques and capital (Bromley 1979).

The relative size of the formal sector varies greatly from city to city. In the poorer countries, the cities contain relatively few large-scale industries and commercial activities; the professionalized middle-class and the blue collar work force are much smaller than in the region's major urban centres. It is difficult to be precise because of the difficulties involved in defining the formal sector but, in major cities such as São Paulo and Mexico City, over half of the labour force are involved in some kind of formal sector activity compared with less than one-quarter in cities such as La Paz and Asunción.

Labour that is not employed in the formal sector is either unemployed or is engaged in some branch of the 'informal sector'. The level of open unemployment is often exaggerated in accounts of the situation in Latin American cities; urban unemployment is generally lower in Latin America than is now characteristic of most developed world countries. It is true that at times of recession levels of unemployment may rise to wholly unacceptable levels in certain cities. For example, in 1982, the level of open unemployment in Santiago was 20.3 per cent, but as Table 7.2 shows this is atypical. The reason for these relatively low unemployment rates, however, is not the widespread availability of well-paid work but the absence of unemployment benefits. An unemployed person cannot register for state benefits in the same way as in the United Kingdom, France or

Table 7.2 Rates of open urban unemployment, 1970–82 (%)

	1970	1978	1980	1982
Argentina	4.9	2.8	2.3	5.7
Brazil	6.5	6.8	6.2	7.7
Colombia	10.6	9.0	9.7	9.3
Chile	4.1	13.3	11.7	20.1
Mexico	7.0	6.9	4.5	3.7
Peru	6.9	8.0	7.1	6.8*
Venezuela	7.8	5.1	6.6	8.2

* 1981 figures

Source: CEPAL, 1982

the United States. Quite simply, the choice for many is between finding some kind of work or going hungry.

In such circumstances, few of the unemployed constitute the poorest members of the society (Berry 1975). To be unemployed a worker requires family support; if a husband is out of work his spouse may work to provide the basic income. So long as an adult is working a grown-up child can wait for suitable work; if a brother is employed his sibling can remain at home. Many of those who are unemployed have technical or educational qualifications; for people with secondary school or university education it may be more sensible to continue searching for a suitable job than to take very low-paid work. If there is a reasonable expectation of obtaining a government job when the current administration hands over power, or of working in a private office because there is a regular turnover of people with the candidate's skills, then waiting will pay dividends. Indeed, for many professional groups in Latin America the irony is that there are more trained workers than jobs. In Colombia, there is a crying need for doctors and yet, because of financial constraints, many trained medical practitioners cannot obtain jobs; they either seek work in other activities, such as business or teaching, or wait for friends with political influence to obtain a post for them.

If many of the unemployed are not among the poorest, this is not to deny that temporary crises may force large numbers of workers out of employment. In general, however, the poor are low-paid rather than unemployed. Most work very long hours for little reward. They are not unemployed or even underemployed but are actively engaged in low productivity work. Fewer people could do the same amount of work, but the paradox is that they are still required to work a long day. Take, for example, the case of numerous street vendors who sell sweets, fruit and cigarettes. Insofar as custom comes regularly through the day, long hours are required; cut the working day and the worker's income will fall. Hence, although sales per worker are limited, the working day is long. Underemployment is obviously not the best term to describe the hours worked by most Latin Americans and the more emotive term 'working poor' is perhaps more appropriate.

Many of the working poor, though by no means all, are engaged in what is often described as the 'informal sector'. This term was originally devised as a means of classifying the occupations of the millions of workers who were excluded from jobs in the formal sector (Hart 1973). The idea was to emphasize the significance of self-employment and small-scale enterprise in providing work. It was subsequently adopted by the International Labour Office which sought to assist the informal sector in an effort to increase both productivity and the number of jobs available.

A basic problem with this concept is the very difficulty of defining the nature of the informal sector. Like most dichotomies, the distinction between the formal and the informal sector is difficult to sustain. While iron and steel plants are clearly part of the formal sector and bootblacks part of the informal sector most activities are less easily classified. The informal sector is especially problematic because it is a residual category for everything excluded from the formal sector. As such it contains a rag-bag of activities with little in common. This imprecision proved a difficulty with respect to policy formulation and indeed to understanding the nature and problems of the informal sector.

Early interpretations of low-wage work had suggested that most jobs were 'involutionary'; the jobs were unproductive because they merely subdivided existing tasks so as to produce work for the unemployed. New jobs were created not because of new needs but because existing functions were subdivided. This categorization of work suggested that informal sector jobs were small-scale, labour-intensive, lacking formally acquired skills and reliant on domestic resources. Most critically, jobs were easy to enter and markets were unregulated by governments. In fact, detailed studies of Latin American cities have revealed the diverse array of jobs contained within the informal sector. Many of these jobs fit uneasily into the classification because they can be entered only with difficulty and often require experience, capital and permits. Indeed, the informal sector appears to be highly structured with a well developed hierarchy of functions and seniority (Peattie 1975). Thus salesmen in the city centre often require licences which have to be bought from the authorities or from former salesmen. Prime sales positions are reserved for those who have been longest in the profession or who can afford to buy the rights to the spot. Bootblacks cannot practise in the best locations without competition with existing workers; cleaning the shoes of clerical staff in an office is the prerogative of a regular bootblack – the porters will not allow a new bootblack into the building. Clearly, the better jobs in the informal sector require experience and contacts. It is no more possible to enter the most remunerative parts of the informal sector than it is to start off as managing director of a formal sector company.

In addition, most of the better paid informal sector activities require capital. Selling lottery tickets requires that a salesman has the money to pay for the tickets in the morning; he is not repaid until he has sold them. Many informal sector workers avoid this constraint by working for someone else, for borrowing money is very expensive. Clearly, ease of entry into better paid parts of the informal sector cannot be achieved without difficulty.

The informal sector does not consist wholly of so-called 'petty services', bootblacks, salesmen, newspaper vendors, beggars, etc. It also comprises the artisan and craft workers, many small garages and repair shops, and other clearly manufacturing activities. Given the common perception that services are 'unnecessary' and manufacturing 'necessary', the discovery that the informal sector also contains the latter helps convince sceptical government officials that the sector contributes to the urban economy. Indeed, more and more literature is contradicting the original idea that the informal sector is in some sense parasitic on the urban economy.

Early work on Latin American cities constantly suggested jobs could be eliminated without any loss in production. These activities were marginal to the main sources of employment and production, and existed only so that people could survive. In addition, it was thought that there was little interaction between the 'productive' and the 'unproductive' sectors. These ideas encouraged governments to provide finance and training for the productive formal sector and to ignore or even discourage the informal sector. The consequences were that many formal sector activities were given easy access to public and private financial agencies whereas small informal sector activities were neglected; many street traders and small businesses were constantly harassed by the police because they lacked licences. So long as the image of the informal sector was one of social rather than economic utility, such policies were likely to endure.

Increasingly, however, this conception of the 'informal' sector is being questioned. During the 1970s the World Employment Programme came to regard the 'informal' sector as a source of opportunities for creating more adequately paid jobs. Advice, credit and new techniques were fed into different kinds of informal activities as a method of creating more labour-intensive jobs. Governments were convinced that harassment of the sector was less appropriate than encouragement.

Other views of the informal sector are also emerging which question additional aspects of our understanding. The belief that the formal and informal sectors operate independently is the most recent casualty. Indeed, a number of writers has argued that profits in the formal sector depend upon the existence of the informal sector. One study in Cali, Colombia, shows how the paper, bottles and waste metal that are collected from the municipal rubbish dump are recycled through the big industrial companies in the city (Birkbeck 1978). This example shows that one of the most 'marginal' of activities is linked to the formal sector, and, because the recycled materials are cheaper than producing new paper, glass or metal, some benefit is clearly derived by the industrial companies from the transaction. Manufacturing companies also benefit in terms of sales from the informal sector. Cigarettes and sweets are sold on most street corners, indeed the sales distribution system could hardly be improved, and yet the companies do not formally employ their sales 'staff'; they do not pay their social security, they pay only for results and laying off staff is no problem. The result for the companies is a cheap, flexible sales force; for the poor, long hours and low wages. The argument can be taken further to argue generally that the cost of labour in the formal sector is reduced by the presence of the informal activities. First, costs of living of factory workers are kept down by the low price charged by shopkeepers and marketsellers operating in the low-income areas of the city. Second, the self-help housing, in which many formal sector employees live, provides cheap accommodation which keeps labour costs down. In this sense, the formal manufacturing sector may be gaining throughout the production and sales process – it obtains higher profits because of cheap inputs and because it obtains a cheap, flexible sales system. In addition, it often persuades governments to act against the informal sector when the latter's activities have ceased to be useful to the formal sector.

Of course, it is erroneous to take this interpretation too far. Low incomes in the informal sector reduce the costs of production and sales of the formal sector, but they also mean that its market is much more limited than would be true if there were a highly paid labour force. How the balance of advantage works out for an individual company is difficult to evaluate; some producers clearly benefit, others lose. For example, companies producing manufactures for export gain twice over; they have a cheap labour force and their market is unconstrained by low domestic wages. By contrast, the domestic market for manufacturers of clothing or refrigerators is limited because of the low incomes of the urban poor.

Whatever the balance of advantage a critical point is that the formal and informal sectors are intimately linked in the Latin American urban economy. The poor are not marginal to the 'productive' sector but form part of it. Their jobs may be generated by the formal sector, but their own activities add to the profitability of the formal sector. Nor are most of the jobs the traditional activities of former peasants. While some craft production persists, most informal activities are modern jobs. Selling American brands of cigarettes, repairing cars, helping shoppers carry their purchases from the supermarkets to their cars, installing electricity or plumbing into self-help homes are not activities that could have existed in 1900. These jobs are not only modern – they are also economically productive and far from 'marginal'.

The pattern of employment that we have just described has its effect on the distribution of income. The fact that many people are employed in low-productivity occupations means that they receive little reward for their work. Many families survive, indeed, only because they have several workers in the household; in São Paulo it has recently been argued the number of workers in each household has risen because of the falling labour incomes (da Camargo 1975). Certainly few workers in Latin American cities earn enough to support a household satisfactorily and where there is only one wage earner major problems ensue. In most cities somewhere around one quarter of children may be undernourished as a result. Whether the general situation is improving or deteriorating is uncertain but it is undoubtedly true that in some cities real manufacturing wages have fallen dramatically during the current recession (Table 7.3).

If many families are poor, there is also a minority of rich families. Most Latin American cities exhibit very wide differences in household incomes. The skills of many professional groups in Latin

Table 7.3 Changes in real incomes, 1978–82 (1970 = 100)

	Construction salary			Industrial salary		
	1978	**1980**	**1982**	**1978**	**1980**	**1982**
Argentina	61	64	59*	72	93	74
Brazil	103	94	96*	127	128	139
Colombia	103	117	121*	91	98	97*
Chile	85	102	96	84	104	103
Ecuador	121	98	91*	135	171	168*
Mexico	134	132	136*	122	115	119
Peru	84	87	94	85	88	86
Venezuela	110	111	104*	119	122	116

* 1981 figures
Source: CEPAL 1982

America guarantee them a good income. If in the formal sector there is no shortage of competition, the relative shortage of engineers, accountants, computer programmers and surveyors, guarantees them rewards that are often higher in purchasing terms than those received by comparable groups in developed countries. In addition, a minority earn large incomes through their ownership of stocks, land and real estate. Thus the distribution of income in urban areas is often extreme, the majority surviving on low incomes with a minority gaining major advantages from their market position or their control of capital.

Social segregation in the city

Latin American cities have clearly segregated areas of land use. There are industrial zones which accommodate modern factories, well developed commercial and rental centres, high-income residential areas, zones of government and private offices, and large swathes of low-income residential development. Some parts of the cities are well ordered and regulated, others lack services and appear to have developed quite spontaneously. This segregation is mainly the result of market forces. Industrial companies and high-income residential groups can afford to bid for high-value land, the poor occupy residual land. The market mechanism is modified by the intervention of governments although most public agencies are forced to buy land like any private company. Governments do influence the price of land through their servicing and planning policies and through taxation. Sometimes, too, governments control large areas of land, a legacy sometimes of colonial rule, as in Venezuela's municipal land, of revolution, as in Mexican *ejidos*, or of the fall of a dictator, as in Nicaragua or Cuba. In general, however, the dynamics of urban land use are determined by the interaction of demand and supply, with only occasional effective interventions by government.

Most Latin American cities have a central core which developed during the colonial period sometimes on the site of an existing settlement, as in Mexico City and Cuzco, but most frequently in a new location. The centre piece of the Spanish American city was the colonial *plaza mayor*, a large square flanked by the cathedral or church, government offices, and other public buildings. The homes of the élite were originally located close to the square along the streets which spread out in a grid-iron pattern. The lower income areas were further out at the edge of the city. Portuguese colonial settlements usually lacked a central plaza, were often less regular in design, and were more likely to follow the dictates of the terrain, even if the differences with the Spanish settlements have often been exaggerated (Gasparini 1981).

Until quite recently, most cities maintained this urban form. It has changed only as a result of changes in housing design, new forms of transportation, the growth of car ownership, and the rapid expansion of the urban population. During the twentieth century, élite groups have gradually moved away from the central city to occupy suburban housing. The growth of large factories has required the reservation of special industrial areas close to railway tracks or main roads. As cars and buses have come to dominate urban travel patterns, new commercial areas have developed along the main routeways between the city centre and major residential areas. Governments have responded by improving road communications, attempting to separate conflicting land uses, and generally regulating the pattern of urban expansion. While many cities still exhibit high population densities by North American standards, there has been a strong movement outwards towards a more dispersed suburban pattern of design, with segregated activity areas.

The dynamics of urban growth are determined by commercial demand and by the market power of high-income residential groups, although some land is reserved for necessary government functions and public services. The land that remains is available for use by those groups which can afford to pay least. Hence the urban poor occupy land that is unpopular among other groups. Such land is liable to flood, suffers from some kind of pollution, or because of its physical characteristics is difficult to service. Whatever the mechanism by which the poor obtain land they occupy the areas which other groups have left unoccupied.

Planning and servicing policy tends to accentuate the patterns of segregation which result. The service agencies first supply areas where the owners are politically powerful and/or can afford to pay the cost of the services; industrial areas and higher income residential areas are always fully serviced. The location of service lines and roads affects the price of land and helps to determine neighbouring land uses. New high-income residential areas will develop close to existing élite areas both to gain access to services and to share the social cachet of the prestige locations. Thus, land values and the way that the land is divided by developers into large or small plots will determine the future patterns of land use. In low-income areas, the lack of adequate services will discourage most higher income groups from moving in. The combination of market forces and servicing policy accentuates social segregation.

The level of segregation can be demonstrated by the maps of Caracas and Bogotá. Both show how high-income and low-income residential areas occupy different sectors of the city. In Caracas, the low-income *rancho* areas congregate on the hillsides, occupying land that has been avoided by other land uses.

In Bogotá, the low-income areas are concentrated in the south of the city and to the north-west. The environmentally more desirable areas to the north are the domain of high-income groups; one poor area located in the north is contiguous to a limestone quarry and cement works. In the south, the land is either liable to flood or is located on hillsides which are difficult and expensive to service.

The housing of the poor

Two stereotypes of poor housing appear in the literature: the one relating to conditions in the central areas of large cities, the other to the peripheral shanty towns. The first picture derives from detailed anthropological observations of conditions in the central cities of Latin America where newly arrived migrants and recent-established native households rent accommodation in crowded multi-family occupancy. Many of the houses were designed for high-income families which have since moved to more modern, suburban accommodation; their former residences having been divided into self-contained rooms to rent. Elsewhere 'purpose-built' accommodation has been constructed for rent. These *conventillo* or *inquilinato* areas contain high population densities and living conditions are poor. One such area in Mexico City, with 700 inhabitants living in a single block, is described by Oscar Lewis (1963: xiv).

> The Casa Grande is a little world of its own, enclosed by high cement walls on the north and south and by rows of shops on the other two sides . . . This section of the city was once home of the underworld, and even today people fear to walk in it late at night. But most of the criminal element has moved away and the majority of residents are poor tradesmen, artisans and workers.
>
> Two narrow, inconspicuous entrances, each with a high gate . . . lead into the *vecindad* on the east and west sides . . . Within the *vecindad* stretch four long, concrete-paved patios or courtyards, about fifteen feet wide. Opening on to the courtyards at regular intervals of about twelve feet, are 157 one-room windowless apartments, each with a barn-red door . . .

183

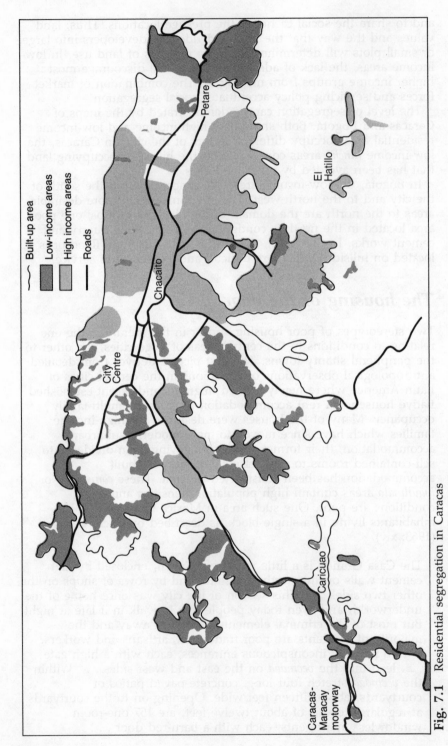

Fig. 7.1 Residential segregation in Caracas

Built-up area

Low-income areas

High income areas

Roads

Petare

El Hatillo

Chacaíto

City Centre

Caricuao

Caracas-Maracay motorway

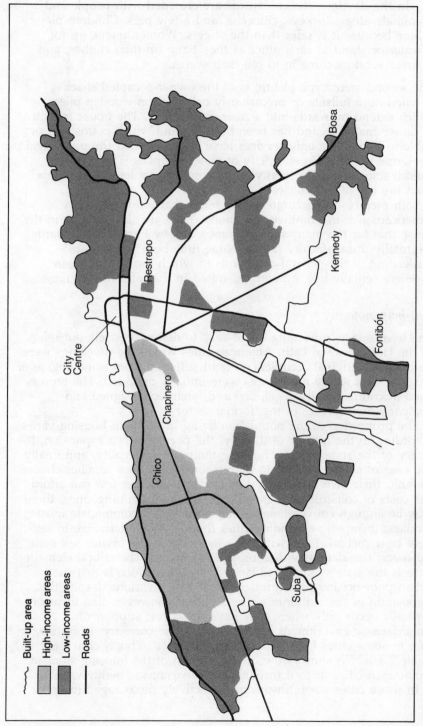

Fig. 7.2 Residential segregation in Bogotá

Built-up area

High-income areas

Low-income areas

Roads

Bosa

Restrepo

Kennedy

Fontibón

City
Centre

Chapinero

Chico

Suba

In the daytime, the courtyards are crowded with people and animals, dogs, turkeys, chickens, and a few pigs. Children play here because it is safer than the streets. Women queue up for water or shout to each other as they hang up their clothes, and street vendors come in to sell their wares.

The second stereotype picture is of the owner-occupied shack perched on a hillside or precariously balanced on wooden piles which extend outwards into a river, sea or lake. The house is built of flimsy materials and has been built on land which is unstable or liable to flood. The only services have been obtained through illegal hook-ups to the main electricity or water supplies. The only conceivable sign of prosperity is the sprouting of television aerials from the top of the houses.

Both pictures are accurate in the sense that too many Latin Americans live in primitive conditions. They are also accurate in the sense that far too many inhabit homes that by European standards are totally unsatisfactory. At the same time both pictures are misleading of the general conditions in which most poor urban dwellers actually live, a point illustrated in the following sections.

Self-help housing

The term self-help housing is not easy to define precisely but refers to the large areas of Latin American cities where the dwellings were built by the original occupiers on land suffering from some degree of illegality and where no services were initially provided. The term is used in contrast to the legal, serviced, architect-designed and company-built houses of the 'formal sector'.

The proportion of the population living in self-help housing varies according to the income of the city, the pace of urban expansion, the policy of the state towards housing tenure and illegality, and finally the ease of access to land. In high-income cities, few families choose to build their own homes; in very low-income cities few can afford the costs of construction materials. In slowly expanding cities, there may be enough conventionally built houses to accommodate most families; in rapidly expanding cities there is little alternative to self-help construction. Even where the lack of housing invites self-help, however, the state may discourage the process. The critical element here is the state's attitude to land. The popular idea is that the urban poor occupy land through large-scale invasions despite the opposition of the authorities. The reality is, however, that invasions typically occur only where they have the covert support of politicians or government administrators. The corollary, of course, is that in some cities invasions are not permitted. This is clear from Table 7.4 which shows how the proportion of the housing stock in Colombian cities formed through land invasions is highly variable.

In those cities where invasions are actively discouraged the poor

Plate 11   In low-income areas electricity supply is unreliable and often illegal. Gradually the service is regularized as the installation of meters in this Mexico City settlement shows. (Photo: A. Gilbert)

Table 7.4 Proportions of urban population living in invasion settlements in selected Colombian cities

Armenia	1.3
Barranquilla	9.1
Bogotá	0.8
Bucaramanga	2.3
Cali	2.8
Cartagena	8.0
Cúcuta	49.9
Ibagué	10.9
Manizales	0.3
Medellín	3.6
Pasto	0.2
Pereira	7.1

Source: ICT 1976

are forced to purchase land. Of course, they cannot afford to buy land in good neighbourhoods and can rarely buy land that is serviced. The only real alternative is to buy land that is illegal in the sense that it has not been sanctioned for urban use by the planning authorities. Such land has not been developed because it lies beyond the urban limits of the city, because it is unserviced, or because the land cannot legally be sold for urban use. The forms of illegal land purchase vary from city to city. On the edge of Mexico City *ejidos* are sold illegally by members of the community to the urban poor. In Bogotá, sales are organized by an illegal subdivider who sells plots of land on the fringe of the city. The formation of the illegal subdivision of Britalia in Bogotá illustrates certain of the features characteristic of the low-income residential land market.

Britalia is located in the south-west of Bogotá on low-lying land which is liable to flooding in winter. It is flanked to the west by open land belonging to a large estate and to the east by other illegal subdivisions. Sales began in 1973 but possession was not granted until 1974 when most lots were in fact sold. The pirate urbanizer bought the land, or at least promised to buy the land from the two owners. By 1976 he had paid off the smaller debt but still owed the bulk of the purchase price. The urbanizer hoped that the settlement would be legalized under the 1972 Minimum Standards Decree but this was not permitted because one-fifth of the planned area lay outside the urban perimeter and because the water company felt it would be difficult to drain the lower parts of the settlement. The request was formally refused in January 1975. In May of that year the Committee for Community Action denounced the urbanizer to the authorities. The latter did not intervene, officially because they lacked confidence in the state authority (ICT) to service the settlement, and signed an agreement with the urbanizer to legalize the community providing he brought the settlement up to minimum standards servicing levels. He complied with part of the request, paid for the installation of electricity, paid off part of his debt to the original owners, and had made considerable efforts to install some water and drainage services. His efforts were not sufficient, however, and in June 1976, urged on by the community, the authorities took over the financial running of the settlement. By 1979 the settlement had around 1,600 families who occupied half of the 2,846 lots available. The mean purchase date of the original households was towards the end of 1974 and the mean moving-in date the beginning of 1976.

In other parts of Latin America, of course, invasions are permitted by the authorities in certain parts of the city. Sometimes the invaders have been covertly organized by powerful personages in the state. In Peru, General Odría, president from 1948 to 1956, encouraged the poor to invade land in Lima as a means of undermining the support traditionally received by the popular APRA party (Collier 1976). Sometimes prior government authorization allowed the land to be occupied during the day with government vehicles carrying the families to the site. Sometimes the link was more covert and the invasion occurred at night. On other occasions opposition political parties organized the invasion which the government was later forced to accept to maintain its public image.

In Venezuela, the state is also heavily involved in the invasion of land as a means of obtaining popular support. Indeed, ever since a dictator in the 1950s destroyed large numbers of *ranchos*, the political parties of the country have encouraged land invasions. While the government may discourage invasions during much of its term of office, under the stress of elections invasions proliferate.

Even in countries where invasions are permitted, however, many efforts to occupy land are repressed. Such repression may occur for a variety of reasons. Sometimes the organizers of the invasion have failed to acquire adequate political support. It is very common, for example, for minor politicians to try to organize invasions to build up a personal following. If they can convince a large group of people to occupy land and the police can be discouraged from removing them, the politicians can count on the backing of the population, especially when they are able to persuade the authorities to provide services. Minor politicians may face opposition from more powerful colleagues, or from members of other political parties. If the police are controlled by the opposition, then the invaders may be removed. If the land that is occupied is a source of embarrassment to powerful political allies the result may be the same. The best recipe is to choose the site carefully: land that belongs to foreigners, especially at a time of national xenophobia, land which is subject to doubts about its ownership, which the owner fails to guard adequately, or which has little commercial value. It is not advisable to occupy land that is contiguous to high-income residential areas or to farms belonging to powerful political families.

The police have been sent in to remove innumerable invasion settlements in Latin American cities. The outcome of their intervention has depended upon the factors just mentioned but also on the nature of the subsequent confrontation. Not infrequently death or injury to some of the invaders has been sufficient to persuade the authorities to recall the police. Where the authorities have been determined, of course, even such a tragedy has had no effect on the outcome. In Caracas, during the 1950s President Pérez Jiménez not only opposed new invasions but demolished large areas of existing shanty housing. The National Guard was used to drive the occupants out of the *ranchos* into newly constructed apartment blocks, a wholly misguided attempt to improve living conditions for the poor (Dwyer 1975; Carlsson 1961).

Once land has been occupied, whether through invasion or through illegal subdivision, the process of settlement development is very similar in most cities. The initial problem is to construct a shelter. Where people have invaded land it is essential to build quickly and to live on the site during construction; in illegal subdivisions construction can proceed at a less frenetic pace. Materials are piled up on the sites and different households build a one or two room house as fast as they can. At first the shelter may be quite flimsy but within a couple of years solid wood or brickbuilt structures will appear. In cities with hot climates, straw or canvas will be replaced by wood or corrugated iron; in highland cities or countries with cold winters brick will be more common. Most families will engage in construction themselves but the extent to which houses are self-constructed will depend on the income of the family, their skills and experience, and the time at their disposal. Few families fail to enlist the help of skilled friends and neighbours for special tasks such as putting in glass, bricklaying, installing electricity points and plumbing; most will pay for specialized help.

Construction of the house is a matter for the individual household, but all families have a common interest in settlement consolidation. Many are likely to belong to, or be involved in, a community organization which petitions for services and organizes communal projects. Electricity and water are the first priorities and the community will approach the authorities and friendly politicians to arrange matters. Where help is not forthcoming or where a long wait is expected, the community may establish illegal electricity lines. Skilled workers will be contracted to acquire the transformers and to link the settlement to the mains. Each family will then pay to be connected to the transformer. Similarly, the settlement may 'splice' into a water main when one passes close to the settlement.

The community and the authorities will negotiate for several years over the provision, legality and costs of services. Large settlements are likely to find favour more quickly especially when elections are close. In Venezuela strange changes of policy occur close to elections; previous prohibitions seem to evaporate, water is supplied to previously unfavoured settlements. Gradually, the neighbourhood receives more and more services and after ten years there is little that distinguishes the self-help settlement from the housing found in most of the city.

The pace of consolidation of course depends upon the income of the families relative to the costs of materials and land (Ward 1976). Where materials are expensive and land has to be purchased the pace of consolidation may be slow. Sometimes it can be very rapid, especially when middle-income families move into the settlement. Once services have been installed and the houses have become consolidated, families will begin to rent out rooms to other families. Frequently, rooms with a separate front door will be constructed. Gradually, the proportion of tenants in the settlements increases as more families rent out accommodation and increase the numbers of rooms they have available to let. Indeed, this process is now so widespread in cities such as Bogotá, that most of the tenant population live not in conventionally built housing but in consolidated self-help areas. As the population of the city has increased, the old central areas have been unable to accommodate the majority of the new households especially as some of these areas have been demolished to make way for new office and commercial developments.

The tenant population

According to the well-known housing model developed by John F C Turner, the tenant population consists mainly of recent migrants to the city who establish themselves in the home of kin or rent accommodation. Once they become established in the city with regular work and perhaps with a growing family they contemplate moving into the urban periphery where they can engage in home construction. We have already suggested that such a move is not simply a matter of choice. Potential home developers need to have accumulated sufficient savings to pay for the cost of land and materials.

Of course, where land can be invaded easily the total cost of home construction and consolidation is likely to be lower than in cities such as Bogotá, Medellín, Mexico City and São Paulo where land is expensive to buy. In such cities it is likely that many people will have to wait a number of years before they can afford to buy land. Indeed, in Bogotá the average age at which home consolidators first obtained a plot of land in the city was 36 years (Gilbert 1983). This figure is reduced to 31 years if adult migrants are excluded, but is still quite an advanced age. It suggests that access to self-help housing is not easy and that many households are forced by their lack of savings to live many years as tenants.

Surveys in two Colombian cities suggest that most tenants are not markedly poorer than owners (Edwards 1982; Gilbert 1983). Both, however, found some families who were confined to rental accommodation for life. These tended to be households with elderly or sick breadwinners, or homes where a woman was the sole income earner. Few families seem to have moved out of ownership into tenancy. This at least supports the direction of tenure change suggested in the Turner model. Indeed, there are suggestions in data from Mexico City that a typical residential history of a migrant or newly formed household changes from sharing with kin, to renting an independent home, to ownership (Gilbert and Ward 1985). The question is less whether people want to own or believe they are able to construct; much more whether they can gain access to the resources required to own.

It would certainly seem that, where land is scarce or where economic circumstances deteriorate, the levels of renting and sharing increase. In Santiago, Chile, the severe economic recession of the 1980s seems to have generated a very large number of shared households, known locally as *allegados*. Throughout Latin America households clearly survive by modifying the form of their accommodation to their economic circumstances. Normally this will mean delaying the move into ownership until they can afford it, but there are circumstances when conditions are so bad that owners will be forced to sell their homes and move into rented or shared accommodation.

Planning for better cities

Urban planning was common in the Americas before the arrival of Columbus. The great Aztec and Inca civilizations produced marvellous ceremonial cities, such as Cuzco and Teotihuacán, planned meticulously by architects and artists under orders from an élite. With the coming of the Spanish and Portuguese, planning changed in nature and scope but continued to dictate the form of the rapidly changing built environment. The hundreds of new cities that were built by the Spanish and Portuguese mostly followed plans based upon the urban designs that had matured in the mother countries. Most urban plans were very rigid and strictly implemented; they laid down the precise shape of the city, its houses and streets; they determined which areas were to be occupied by different social groups.

The common link between pre-Columbian and colonial urban planning is that the rules were dictated by an élite with little regard for the effect on the poor (Hardoy 1982). Indeed, providing that the broad urban design was not affected, the organization of the poor's housing was a matter of little importance to the planners. Arguably, independence and the many changes effected during the ninetenth and twentieth centuries, have brought little change to planning in this respect. Urban plans are still concerned primarily with the ceremonial and the large scale, with the needs of the rich and the powerful. Even today the realm of the professional architect and planner hardly intrudes into the urban design of the poorer zones. As Hardoy argues: 'Contemporary Latin American cities, like their predecessors, have been constructed by many builders, mostly anonymous, forced to find their own solutions in the face of government neglect and sometimes repression.' This view contradicts, of course, the conventional wisdom of the past fifty years. Urban planning has become a world-wide practice on the premise that it brings benefits to all groups in society. In some respects this premise has been substantiated, for example, through improvements to electricity, water, sewerage and transport networks and the more generalized access to these services that urban growth has brought. The general claims, however, that urban planning benefits all social groups and improves the structure of urban society have not materialized. Good and fair planning should help most urban groups; current practice too often fails to achieve this goal.

There are many signs, indeed, that planning has been introduced not because it would create a better human environment but because it could be used as an additional weapon in the armoury of the powerful. Let us consider, for example, the effect that urban planning has had on social segregation. Much of the literature on urban planning has argued that different social groups should live within the same neighbourhoods. Many urban planners have acted as social engineers striving to reduce inequalities in society. Within urban Latin America, however, practice has been different; there has been a '. . . deepening of class divisions within these cities, reflected in increasingly well-defined patterns of spatial segregation by social class' (Cornelius 1977). Of course, such segregation might have been more acute without the intervention of planners. It is possible that residential differentiation was inevitable given such wide differences in per capita wealth and income and that urban planning has simply made little difference, but there are also depressing examples where planning may actually have accentuated the level of segregation. As Cornelius notes: 'most government interventions (public housing projects, urban renewal schemes, zoning regulations, investments in urban infrastructure, and provision of public services) have only reinforced the market forces promoting class segregation . . . In fact, most governments seem intent upon "building the divided city".'

Specific cases where government intervention has accentuated residential segregation are easy to find. In Bogotá, practically all government housing schemes have encouraged the natural tendency towards class segregation. Urban land-use zoning may have had a similar effect. Indeed, Amato (1968) argues that such zoning was introduced into the city to maintain segregation. It was introduced not with reformist zeal but as a response to alarm among the élite that 'their new residential areas might be undermined by the encroachment of undesirable uses' at a time of rapid urban growth. He claims that it achieved its purpose: 'the early zoning of the city permanently fixed the spatial distribution of socio-economic groups and bears testimony to the desire and ability of the élites to structure the city in conformity to their own interests'. If this view overstresses the influence of élites in the formulation of the zoning scheme it shows how zoning helped these groups insofar as it protected their residential areas.

Such negative examples merely show that urban planning is a limited process that is incapable of modifying the social structure of the power system. If it were introduced into an equitable society in which all social groups participate in decision-making, its effects on the urban fabric would be very different. The reality, however, is different in most of Latin America where authoritarian regimes rule in the interest of limited groups of the populace. In such circumstances it is perhaps not surprising that planning has often been misused. For example, urban planning has often been adopted only as a means of obtaining foreign loans, a practice tied to the trend in recent years for agencies such as the World Bank to insist that loans be linked to rational and technocratic forms of decision-making. As such, some governments have embraced urban planning not as a means of restructuring urban society but as a mechanism for obtaining hard currency credit; planners have been used merely to dress government projects in suitable clothes to parade before the international lending institutions.

Of course, numerous efforts have been made to use urban planning in the redesign of urban society. Unfortunately, few of these efforts have got very far for they have suffered from a variety of difficulties. First, many cities have grown beyond their original administrative boundary and now occupy several political-administrative fragmentation hinders effective land-use control, expanded beyond their original municipal boundaries. Such administrative fragmentation hinders effective land use control, public service delivery and most kinds of forward planning. Second, Latin American cities seldom have sufficient resources to finance the costs of services and infrastructure. Attempts to raise taxes are often blocked by the rich and powerful and, without an adequate tax base, essential services cannot be provided for the poor; health, education and social welfare facilities are unsatisfactory for most low-income groups. Third, the very nature of government administration in Latin America undermines many efforts to plan urban development sensibly. High-income residential developments often ignore the planning regulations, usually without sanction. Bribes or influence deflect the attention of the regulatory agency or convince the planning authority to change the regulations. Low-income settlements 'splice' into the main water pipes and are protected by a political patron. In Latin America, the line between partisan politics and administration is much more blurred than in the United Kingdom.

In short, urban planning fails to resolve the problems of urban society. It fails both because urban problems are complex and pervasive, and because planning mechanisms reflect the balance of power in Latin American cities. Since the state is seldom democratic, state agencies rarely act in the interests of the whole society. Sometimes, agencies directly assist the poor, but such interventions are rare. Sometimes, the poor gain indirectly from the planning interventions which are intended to help other groups; expansions of the water supply needed by industry may bring improvements to domestic water provision. But generally urban planning is merely one more mechanism by which the rich and powerful manipulate decisions to their own benefit; a different form of urban planning can only be implemented in a different form of society.

Further Reading

Amato P (1968) *An analysis of the changing patterns of élite residential areas in Bogotá, Colombia.* Dissertation Series No. 7. Cornell University, Latin American Studies Program, Ithaca, N Y.

Amato P (1970) 'Elitism and settlement patterns in the Latin American city', *Journal of the American Institute of Planners* **36**, 96–105.

Balán J, Browning H and **Jelín E** (1973) *Men in a Developing Society.* University of Texas Press. (A detailed and perceptive account of migration to the Mexican city of Monterrey. Probably the best account of migration in Latin America yet written.)

Berry R A (1975) 'Open unemployment as a social problem in urban Colombia: myth and reality', *Economic Development and Cultural Change* **23**, 276–91.

Birkbeck C (1978) 'Self-employed proletarians in an informal factory: the case of Cali's garbage dump', *World Development* **6**, 1173–85.

Bromley R (1979) *The Urban Informal Sector: Critical Perspectives on Employment and Housing Policies.* Pergamon Press, Oxford.

Bromley R and **Gerry C** (1979) *Casual Work and Poverty in Third World Cities.* John Wiley, Chichester. (Includes several chapters about the employment situation in Latin American cities. Discusses the relationships between poor urban workers and the 'formal', large-scale sector.)

Butterworth D and **Chance J K** (1981) *Latin American Urbanization.* CUP. (Has strong sections on migration to the city and the adaptation of migrants to the urban environment.)

Carlson E (1961) 'Evaluation of housing projects and programmes: a case report from Venezuela', *Town Planning Review* **31**, 187–209.

Castells M (1983) *The City and the Grassroots.* Edward Arnold, London. (Includes four chapters on Latin America which presents the latest thinking of one of the most influential writers on urban matters. It is also one of his more clearly written contributions.)

Centro de Estudios Urbanos (1977) *La Intervención del Estado y el Problema de la Vivienda: Ciudad Guayana*, Caracas.

CEPAL (1982) *Notas sobre la economía y el desarrollo de América Latina*, No. 372 Nov/Dec 1982.

Collier D (1976) *Squatters and Oligarchs, Authoritarian Rule and Policy Change in Peru.* Johns Hopkins University Press, Baltimore. (Account of the policy of successive Peruvian governments to the creation of barriadas in Lima. An interesting and provocative account.)

Colombia, Departamento Administrativo Nacional de Estadística (1977) *La población en Colombia 1973: Bogotá.*

Colombia, Departamento Administrativo Nacional de Estadística (1978) 'Alquileres de vivienda en cinco ciudades 1977', *Boletin Mensual de Estadística*, No 328.

Cornelius W A and **Kemper R V** (1977) 'Metropolitan Latin America', *Latin American Urban Research* **6**. Sage Publications, London. (Contains individual descriptions of current problems facing nine metropolitan centres in Latin America.)

Cornelius W A and **Trueblood F M** (1975) *Latin American Urban Research* **5**. Sage Publications, London.

da Camargo C P *et al.* (1975) *São Paulo: Crescimento e Pobreza.* Ediçoes Loyola, São Paulo.

Dwyer D (1975) *People and Housing in Third World Cities.* Longman, London.

Edwards M A (1982) 'Cities of tenants: renting among the urban poor in Latin America', in **Gilbert A G, Hardoy J E and Ramírez R** (eds) *Urbanization in Contemporary Latin America*, pp 129–158. John Wiley, London.

Fox R W (1975) *Urban Population Growth Trends in Latin America.* InterAmerican Development Bank, Washington DC. (Contains statistical data on the development of urbanization in the six major Latin American countries, including the growth of individual cities.)

Gasparini G (1981) 'The present significance of the architecture of the past', in **Segré R** (ed.), *Latin America in its Architecture.* Ch. 3. pp. 77–104 Holmes & Meier Publishers, New York.

Gilbert A G (1983) 'The tenants of self-help housing: choice and constraint in the housing market', *Development and Change* **14**, 449–477.

Gilbert A G and **Goodman D E** (1976) 'Regional income disparities and economic development: a critique', in **Gilbert A G** (ed.) *Development Planning and Spatial Structure*, Ch. 6. pp. 113–42. John Wiley, Chichester.

Gilbert A G and **Ward P M** (1985) *Housing, the State and the Poor: Policy and Practice in Three Latin American Cities.* CUP, Cambridge. (Detailed case studies of the land and housing markets in Bogotá, Colombia, Mexico City, and Valencia, Venezuela. Discussion of the role of the state and public participation in housing and servicing the poor.)

Gilbert A G, Hardoy J E and **Ramírez R** (eds.) (1982) *Urbanization in Contemporary Latin America.* John Wiley, Chichester. (Contains accounts of the housing situation in Brazil, Colombia, Ecuador, Mexico, and Venezuela. Also contains chapters on the history of urban development in Latin America, on employment problems in the region, and on urban planning and renewal in Brazil.)

Gwynne R N (1978) 'Industrial development in the periphery: the motor vehicle industry in Chile', *Bulletin of the Society for Latin American Studies*, No. 29, 47–69.

Hardoy J E (1982) 'The building of Latin American cities', in **Gilbert A G, Hardoy J E** and **Ramírez R** (eds.) *Urbanization in Contemporary Latin America*, Ch. 2, pp. 19–34. John Wiley, Chichester.

Hart J K (1973) 'Informal income opportunities and urban employment in Ghana', *Journal of Modern African Studies* **11**, 61–89.

ICT (1976) *Inventario de Zonas Subnormales de Vivienda, 1976* Bogotá.

Lewis O (1963) *The Children of Sánchez: Autobiography of a Mexican Family.* Vintage Books, New York.

Lloyd P (1981) *The 'Young Towns' of Lima.* CUP Cambridge. (Contains a useful resumé of the housing situation in Peru and a case study of a small settlement in the city.)

Peattie L R (1975) 'Tertiarization and urban poverty in Latin America', *Latin American Urban Research* **5**, 109–123.

Peattie L R (1979) 'Housing policy in developing countries: two puzzles', *World Development* **7**, 1017–22.

Roberts B (1978) *Cities of Peasants.* Edward Arnold, London. (A sociological analysis of ubanization in Latin America which is especially useful for its accounts of employment and of the emergence of urban systems during the nineteenth and twentieth centuries.)

Travieso F (1972) *Ciudad, Región y Subdesarrollo.* Fondo Editorial Común, Caracas.

Unikel L (1976) *El Desarrollo Urbano de México: Diagnóstico e Implicaciones Futuras.* El Colegio de México, México.

Ward P M (1976) 'The squatter settlement as slum or housing solution: evidence from Mexico City', *Land Economics* **52**, 330–46.

Ward P M (ed.) (1982) *Self-help Housing: A Critique*. Mansell, London.

Wilkie J W and **Perkal A** (eds.) (1983) *Statistical Abstract of Latin America*, **22**. UCLA, Latin American Centre Publications, Los Angeles.

World Bank (1973) *World Development Report 1983*. Washington DC.

CHAPTER 8

Rural change: progress for whom?

Janet Townsend

A generation of growth?

In Europe and the Americas, rural areas and rural people in Latin America are often depicted by the media as stagnant, isolated, traditional, resistant to change. These images tally with the real patterns of population change across the continent, with the declining importance of agricultural production as compared with industry, with the slow growth of food production per head, with the recently rising food imports of such countries as Mexico and Brazil, and above all with the persistence of poverty in rural areas (Table 8.1). All this can be very misleading, for the underlying realities are very different.

What growth?

Over the last generation, rural Latin America has seen some of the biggest changes since the Iberian conquests. As a result mainly of new technologies and altered tenures, agricultural production grew from 1960 to 1980 at an average 3.1 per cent per year, slightly faster than the population, including city dwellers. Agriculture increased in area and intensity. Between 1950 and 1975 the agricultural area doubled, while the irrigated area expanded by 50 per cent, tractor numbers by 550 per cent and fertilizer use per hectare by 720 per cent. New seeds came with the new technologies. In 1979, Mexico's average yield of wheat per hectare was 50 per cent higher than that of the United States; Mexico's overall wheat production had quadrupled since 1950. Latin America produced twice as much wheat in 1979 as in 1950, three times as much maize and more than three times as much rice. Nevertheless, it is still estimated that almost 20 per cent of deaths in Latin America are caused by malnutrition: agricultural growth has been no panacea. There has been little expansion in the potatoes, manioc and beans which are the staple foods of the poor.

Agriculture has not, like Topsy, just 'grow'd'. Innovation has been

199

Table 8.1 Rural change and rural indicators

Country (in ascending order of income per head, 1980)	Rural population (percentage of population in rural areas and towns under 20,000)			Agricultural production (percentage of GDP)	
	1950	1960	1980	1960	1980
Haiti	85	92	87	n.d.	n.d.
Honduras	93	89	80	37	31
Bolivia	80	77	73	26	18
El Salvador	87	82	80	32	27
Nicaragua	85	77	69	24	23
Peru	82	72	60	18	8
Guatemala	89	85	84	n.d.	n.d.
Dominican Republic	89	82	70	27	18
Colombia	77	63	54	34	28
Ecuador	82	72	65	29	13
Paraguay	85	84	79	36	30
Cuba	64	61	57	n.d.	n.d.
Costa Rica	82	76	73	26	17
Panama	77	67	71	23	n.d.
Brazil	80	72	61	16	10
Mexico	76	71	65	16	10
Chile	57	49	39	10	7
Argentina	50	41	34	16	13
Uruguay	47	39	35	19	10
Venezuela	69	53	41	6	6
United Kingdom	n.d.	n.d.	n.d.	4	2
United States	49	42	36	4	3

Sources: World Development Report, Statistical Abstract of Latin America, FAO Production Yearbook, Statistical Yearbook for Latin America.

markedly disparate and fragmented, responding in part to dissimilarities in physical conditions, structures of production and market possibilities, and in part to state intervention (see Ch. 6) through pricing, credit and tax policies and through large investments in commercial infrastructure, particularly roads. The road revolution above all has changed the face of rural Latin America, promoting commercialization of agriculture in very dissimilar areas, tropical and temperate, high and low, even settled and 'frontier'. In the tropical rain forest, people will clear and farm new land because they hope that there will be a road. At first, they will have to carry their harvest to a market on their backs; but, if a road comes, there will be an assured profit. In the central Andes, in contrast, peasants may have been growing most of their own food

Table 8.1 (cont.)

Agricultural employment (percentage of economically active population		Agricultural growth (1979 output; 1966 = 100)	Illiteracy (percentage of adults illiterate, early 1970s)		
1960	1980		Urban	Rural	Total
80	74	122			77
70	63	152	21	54	
61	50	166	16	53	
62	50	164			38
62	39	162	20	65	
52	40	117	13	51	
67	n.d.	175	28	69	
67	49	146	19	43	
51	26	174	11	35	
58	52	148	10	38	
56	49	177	11	26	
39	23	158			4
51	29	169	5	17	
51	n.d.	170	6	38	
52	30	159	14	41	
55	36	140	17	40	
30	19	107	7	27	
20	13	148			7
21	11	100	5	11	
35	18	159			24
4	2	123			1
7	2	135			1

and selling a little which will pay the rent and buy salt and tools, once it has reached the market expensively on muleback. If a road comes, these same peasants may buy fertilizers and concentrate on producing a single crop, perhaps potatoes, for the distant city, thus changing their whole agro-ecosystem. For, surprisingly enough, modern road transport is much cheaper per ton-mile than cross-country transport by ox-cart, mule, canoe or even one's own back. The road revolution and the greatly increased commercialization of agriculture create new activities, new jobs, in the cities, in the towns, even in the villages; whole new settlements have appeared, not only on the frontier of settlement but in traditional peasant areas. Direct employment in agriculture, on the other hand, may actually fall.

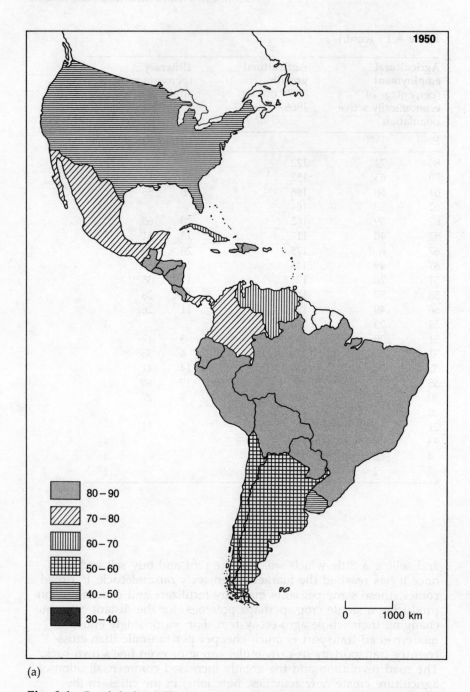

Fig. 8.1 Rural decline? The percentage of population in rural areas and in towns of under 20,000 inhabitants (a) 1950 (b) 1980

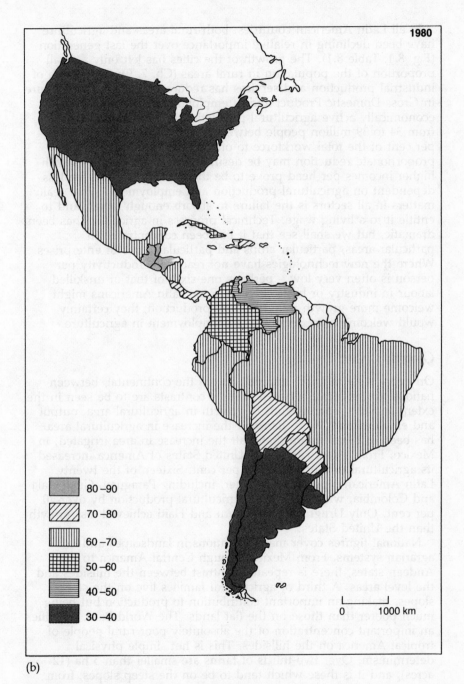

1980

80 – 90

70 – 80

60 – 70

50 – 60

40 – 50

30 – 40

0   1000 km

(b)

Fig. 8.1 (cont.)

In all Latin American countries, both rural areas and agriculture have been declining in relative importance over the last generation (Fig. 8.1; Table 8.1). The growth of the cities has left only a small proportion of the population in rural areas (Ch. 7, 9); the growth of industrial production and services has reduced the role of agriculture in Gross Domestic Product and in employment. Although the economically active agricultural population in Latin America rose from 34 to 39 million people between 1963 and 1980, it fell from 54 per cent of the total workforce to only 35 per cent. This proportionate reduction may be desirable, since the countries with higher incomes per head prove to be the more urban and the less dependent on agricultural production and employment. What really matters in all sectors is the failure to absorb enough labour and to entitle it to a living wage. Technical progress in agriculture has been dramatic, but we shall see that it has been concentrated in particular areas, particular crops and particular types of enterprises. Where the new technologies have not reached, productivity per person is often very low – perhaps one-sixth of that of unskilled labour in industry or basic services. Poor Latin Americans might welcome more growth in agricultural production; they certainly would welcome more remunerative employment in agriculture.

Growth where?

On every scale in Latin America – from the continental, between nations, to the local, between farms – contrasts are to be seen in the extent and the repercussions of growth in agricultural area, output and employment. More than half the increase in agricultural area has been in Brazil, more than half the increase in area irrigated, in Mexico. From 1966 to 1979, the United States of America increased its agricultural production by 35 per cent. Sixteen of the twenty Latin American republics did better, including Paraguay, Guatemala and Colombia, which increased agricultural production by over 70 per cent. Only Uruguay, Chile, Peru and Haiti achieved less growth than the United States.

National figures cover many variations in landscapes and in agrarian systems. From Mexico through Central America to the Andean states, there is repeated contrast between the hillsides and the level areas. A third of agricultural families live on the steep slopes, making an important contribution to production but being much poorer than those on the flat lands. The World Bank identifies an important concentration of the absolutely poor rural people of tropical America on the hillsides. This is not simple physical determinism. Over two-thirds of farms are smaller than 5 ha (12 acres), and it is these which tend to be on the steep slopes; from Mexico to Bolivia, an increasing number of farm families have been attempting to subsist on farms that, over time, have become progressively smaller but little more productive. Often it is on the adjoining flat lands that agricultural production has increased enormously, frequently drawing seasonal labour from the hillsides. Sometimes, as in Mexico or north-west Argentina, irrigation has played a major role in increasing crop yields on the flatter lands; elsewhere it has been other technical changes, such as mechanization, fertilizer, pesticides or improved seeds. Of the industrial and export crops, most come primarily from the flatter lands: bananas, cotton, export sugar, oilseeds, soya, cacao and most fruits and vegetables for canning and freezing. The one export crop which comes mainly from the slopes is coffee. Most locally consumed foodstuffs, on the other hand, are produced primarily on the slopes – potatoes, wheat, manioc, beans, fresh fruits, household sugar and most maize and plantains; only rice is a dominantly lowland product. Industrial and export crops have attracted the advanced technology, while food crops for the domestic market tend to be labour-intensive, and to be produced on tiny holdings on difficult terrain.

In a single country, Colombia, from 1950 to 1976, production of new commercial crops such as cotton, sugar, rice, soya, sorghum, barley and oilseeds increased by almost 600 per cent. Over the same period coffee production increased by only 50 per cent but other traditional crops, for export and domestic use, by 150 per cent. The wage economy expanded with production: wage labourers came to make up half the economically active rural population, and many smallholders also depended in part on wage labour. Here, side by side, were the growth, the commercialization and the deprivation – for malnutrition continued, particularly among women and children in the families of landless agricultural labourers. As it happened, many farmers on steep slopes in Colombia in 1976 were about to make windfall gains from frost damage to Brazilian coffee and from an unexpected opportunity to profit from growing marijuana. Their new prosperity, however much evidenced by new homes and schools, is probably transient; so is that of the Chapare region of Bolivia, where coca has brought similarly unexpected gains. Neither other people's frosts nor the production of illicit drugs are a secure base for the future.

From Mexico to Chile and from Argentina to north-east Brazil, the 'latifundia-minifundia syndrome' recurs in the landscapes: large estates occupy the best land and tiny farms are crowded together (Table 8.2), often on land which is so steep, or is so susceptible to erosion, flood or drought, that it would be better left uncultivated. Usually the physical contrasts are marked, but sometimes land tenure may be the sole explanation of land use. Latifundia landscapes, in the Andes or north-east Brazil, may be striped by tiny holdings along the roads and railways. The traveller who observes plantains growing along a railway through Colombian ranches, or maize and beans by the road in north-east Brazil with cotton or pasture behind, may think this is due to variation in soil. In reality, access to land by the small producers may be so difficult that they are reduced to cultivating the railway embankment or the roadside

Table 8.2 North-east Brazil 1972:
inequality in distribution of land

Farm size (ha)	Percentage of farms	Percentage of land
0–9.9	32.0	1.4
10–49.9	35.1	9.1
50–99.9	12.5	8.7
100–199.9	9.8	13.5
200–499.9	6.4	18.7
over 500	4.2	48.6

Kutcher and Scandizzo have shown
that these discrepancies cannot be
explained by land quality, location
or value.

Source: Kutcher and Scandizzo (1981)

verge because, as they will say, 'There is nowhere else to sow a
seed'. Traditionally, the large estates were relatively unproductive,
usually achieving far lower yields per hectare than the nearby
smallholdings. Now, with agricultural growth, many estates have
adopted technological advance and have become highly productive.
Small farmers tend to maximize use of labour, while the estates
are capital-intensive; crops, as we have seen, tend to differ.

The experience of improved varieties of rice in Colombia illustrates
how technical advances may increase the difference between regions
and between farmers, even perhaps acting to the disadvantage of
the poor. Until 1967, more than 90 per cent of Colombian rice
farmers used varieties developed in the United States, which could
hardly give of their best under tropical conditions. Colombian rice
research then developed varieties capable of very high yields under
irrigation in Colombia: by 1973, all irrigated rice land was planted
with them. The average yield for irrigated rice increased from 3 tons/ha
(2½ acres) in 1966 to 5.3 tons in 1973, while yields on unirrigated
land remained at 1.5 tons. Meanwhile, the overall increase in output on
irrigated land brought the price of rice down sharply, by 28 per cent;
only the farmers with the new varieties, on irrigated land, could
reduce their costs enough to keep production profitable. The area
under unirrigated rice shrank from 50 per cent to 9 per cent of the
total rice area; farmers without irrigation lost heavily and were
deprived of what had been a profitable crop. When a new
technology is developed, the distribution of benefits among farmers
will be greatly influenced by the pre-existing distribution of control
over the different types of land. In the Valle del Cauca region, high-
yielding hybrid maize was introduced on almost all maize-producing
farms and average yields rose to three or four times the national
average. They had been selected to perform well in Valle del Cauca;
elsewhere, the hybrids were introduced on very few farms and their

yield was much lower. In addition, farmers are faced with the necessity of buying fresh hybrid seed every season; small farmers with poor access to market have always depended on producing their own seed. Large maize producers in Valle del Cauca have thus benefited from the hybrids, while small producers elsewhere have not. It is natural that governments should want agricultural research to increase production and to lower costs. It is then logical to seek improved varieties suitable for those environments considered optimal for the crop. The outcome of the research may discriminate against other regions, or against those without access to new inputs or to certain types of land. In the case of a staple crop such as maize grown by many farmers, only a small proportion will benefit from higher-yielding varieties. In new programmes for improved manioc (a widespread staple of the poorest), Colombian researches are having to design new forms of commercialization in the endeavour to reach poorer farmers. The manioc must become marketable, as animal feed or industrial 'chips' or starch, or it will remain beyond their reach.

Growth for whom?

Clearly, rural poverty (Table 8.3) cannot be attributed solely to a lack of change. Agricultural growth and rural change may work against some regions, farmers and people. Despite growth and real technical achievement, Latin America today fails to produce an adequate livelihood for many rural people. Land and work are difficult to

Table 8.3 Approximate incidence of poverty c. 1970

	Percentage of households poor*		Percentage of households destitute[+]	
	Urban	Rural	Urban	Rural
Honduras	40	75	15	57
Peru	28	68	8	39
Colombia	38	54	14	23
Costa Rica	15	30	5	7
Brazil	35	73	15	42
Mexico	20	49	6	18
Chile	12	25	3	11
Argentina	5	19	1	1
Venezuela	20	36	6	19
Latin America	26	62	10	34

* Without access to minimum goods and services
+ Without access even to a minimal diet
Source: Altimir (1982)

obtain and frequently very insecure. In rural Brazil, there are 15 million migrant labourers or 'cold lunch people' who may have to seek a new but probably still temporary job every few days or every few weeks. Rural demands for labour are not high, and long-term employment is hard to find. Where agricultural production has greatly increased, there have often been labour-saving innovations such as mechanization, fertilizer and pesticide, which have the effect of exaggerating the seasonality of the periods of peak labour demand. Wages in agriculture are proportionately lower than in cities. This is somewhat offset by cheaper housing: in Brazil, malnutrition has been shown to be more widespread among the urban than the rural poor, apparently because of the higher proportion of wages spent on housing. Other primary necessities, such as clean water and basic sanitation, are less often available in rural areas. The majority of urban Latin Americans have access to piped water, but in rural areas only a handful have satisfactory drinking water. Not surprisingly, most poor rural people suffer not only from infectious diseases and intestinal parasites but from lack of medical care. In Colombia there are 1,000 people to every doctor in the towns, but 6,400 people in rural areas. Most Latin American countries now have primary health care but little is spent on it. Commercial drug networks do penetrate the rural areas, but provide only very poor advice to the consumer, and make heavy demands on rural incomes.

The very means for betterment are less obtainable in rural areas. Education in most of Latin America is strongly linked to income. The rural population is hardly qualified to earn high incomes (Table 8.1): rural illiteracy is usually two or three times as high as in towns. Urban schooling is cheaper and easier to provide, but the successful literacy crusades in Cuba (1961) and Nicaragua (1980) suggest that rural literacy is possible, even for a poor population. Beyond that, even when rural education levels are rising, they rise even faster in urban areas.

Poor rural households can be divided into two groups: those which have access to land and those which are landless. These groups can overlap: many poor households have multiple sources of income, perhaps including urban remittances, and many individuals engage in different activities over time. Perhaps a third of rural households are landless, but many peasant farmers are even poorer than landless workers. In Latin America some 62 per cent of the rural population lack the means to satisfy their basic needs, and 34 per cent cannot even afford the minimal diet necessary to health. In many areas change has worked against both the landed and the landless poor. North-east Brazil is an example: it has almost 40 per cent of Latin America's very poor rural people. Here, for at least twenty years, inequality of incomes in rural areas has been widening; production has risen and technology advanced, but the bulk of the extra income has accrued to those above the poverty line. Here, as elsewhere in Latin America, it is possible to have statistics about people dominated by poverty while statistics of production show growth. In the landscape, the very concentration of wealth leads to an impression of extraordinary dynamism in many small towns, where consumer goods are vigorously traded. Many of the 38 per cent of rural people who can satisfy their basic needs can indeed display the growth and prosperity which is the other side of the coin to Latin American rural poverty.

The conditions for change

Rural change in Latin America is not independent of the 'Latin American World' (Ch. 1). All Latin American economies are deeply 'dependent': their structure and development are strongly conditioned by structures and events in the advanced industrial economies of the world. Dependence on manufactured imports has been reduced but Latin American countries have become more open to the world market in general. From 1970 to 1980, the value of their trade (imports plus exports) as a percentage of GDP rose from 26 to 50 per cent. This international trade, with foreign direct investment and loans from foreign commercial banks, has tied rural as well as urban Latin America more closely to the world market. The process started with the Iberian conquests (Chs. 2 and 3), and accelerated in the nineteenth century when the independent states became primary exporters. More recently, more rural people have become purchasers of consumer goods, and agriculture has become much more commercialized. In many places, landscapes and social relations were redrawn after Conquest and Independence, and are now being redrawn again.

Transnationalization

Despite the relative decline in the importance of agriculture, the involvement of rural Latin America in the world economy has become more and more direct. Although the continent has lost much of its importance as an exporter of food and the percentage of exports derived from agriculture has declined, absolute quantities of agricultural exports continue to rise and new exports are constantly sought, as witness Chile's success with fruit and vegetables, Brazil's with frozen chickens and orange juice, and Colombia's with cut flowers. Conversely, a decision to import cheap food may affect farming patterns nationwide: imports of wheat into Colombia have reduced national production by two-thirds since 1950.

Only a tiny fraction of Latin American rural people remain purely subsistence cultivators who produce neither goods nor labour for the national economy. Most depend on the market not only for selling their products but for buying essential farm inputs. This commercialization has been promoted not only by new roads but by high energy technology. Fossil fuels concern everyone:

mechanization may be a feature of large farms, but pesticides, fertilizers and herbicides interest the smallest producers. Growth in agricultural production has been achieved not by developing locally available knowledge and skills (shown in Ch. 3 to have immense potential) but by agrarian systems which are based on cheap fossil fuels for heavy use of energy and chemicals. It happened that from the Second World War to 1973 the real price of oil was falling. At that very time in Latin America, new transport networks were being planned, the economies were growing apace and high energy technology was reaching large and small producers. Since the rise in the price of oil in 1973, Latin American states have been struggling with price policies which affect even poor and remote rural producers. Agricultural production of food, industrial raw materials and exports call for cheap inputs, transported by cheap fuel – but world prices of inputs and fuel have risen massively. From the state to the smallest producer, everyone was trying to adjust. Then came world recession, the decline in export prices and the crushing burden of debt, which seemed to make adjustment impossible.

High energy technology reached Latin American agriculture with much assistance from transnational corporations. Pesticides are an example of sophisticated chemicals which have costs as well as benefits. In the absence of a skilled, independent advisory service, farmers tend to over-use pesticides; this raises their costs and risks both the elimination of beneficial species and the appearance of resistant strains of pests. Side-effects are equally alarming. When Central American beef destined for the USA is checked for pesticide residues, significant quantities are rejected as dangerous: the cattle have grazed on forage contaminated by pesticides used on cotton. People working on the cotton are even more at risk; they lack protective clothing, and pesticide poisoning is frequent and sometimes fatal. Excessive use of chemicals to kill insects promotes resistance, which can affect not only agricultural pests but human pests such as malarial mosquitoes. Agricultural insecticides have thus played a major role in the survival of malaria when the health authorities had hoped to eradicate it. Resistance to insecticides rarely arises from the planned, limited spraying designed by the health authorities for residential areas; much more commonly, it is a reaction to similar insecticides applied for agricultural purposes. (The World Health Organization considered calling for a ban on such insecticides, but concluded that the effect on crop production would be so severe that the ban would be unacceptable.) Finally, many pesticides sold in Latin America are so toxic that they are banned, or their use is severely restricted, in the rich industrial countries which make them. The United States Agency for International Development was required by the US courts to examine the impact, costs and benefits of its loans. The controls which it now implements in loans for pesticides would, if they were widely adopted in Latin America, change the whole pattern of pesticide use. That is highly unlikely; so is adequate independent advice for farmers. Meanwhile, highly toxic products are sold with the most improper advertising and labelling. The benefits accrue primarily to farmers with ample access to other inputs; the costs are borne by their labourers, by illiterates who cannot read any warnings on the packets, and by those who lack ready access to clean water and health care: the poor.

High energy technology has not been taken up everywhere to the same extent. As the nature of Latin America's involvement in the world economy changes, so the possible roles for specific rural areas alter. When Latin America was primarily a source of raw materials for the industrial economies, rural areas had their importance in production, whether of the raw materials, of food for the national market or of labour for other areas. Now that these countries are more urban and some have become significant world producers of industrial goods, there are extensive national markets for rural products and rural areas have simultaneously become significant markets for urban-industrial goods and services. Above all, they are a market for agricultural inputs.

High energy technology in agriculture varies greatly from crop to crop and from place to place; penetration and profitability are highly uneven. In Guatemala, aerial crop-dusters may be treating cotton close to labourers planting maize with a digging stick; in north-east Brazil, hourly-paid women may apply handfuls of fertilizer to tomatoes irrigated by electric pumps; in the Andes, chemical fertilizer may be ploughed in by oxen. The state itself may promote the unequal use of technology. In Brazil, more than two-thirds of all rural credit goes to the south and south-east; more than half is used as large loans by large farmers; most goes to coffee, sugar, rice, maize, soya and wheat, and a fifth is spent directly on fertilizers and pesticides. Across Latin America, credit and prices are much influenced by the state, and there is heavy discrimination between crops, regions and sizes of farms; this affects the way in which the comparative advantages of regions, of local areas, of farms change over time. Brazil, in its search for non-traditional exports, increased its exports of soya products from a value of $15 million in 1965 to $1,644 million in 1979: this transformed whole regions of the south-east, where traditional family farms have become part of highly capitalized co-operatives selling produce to transnational corporations. Many small farmers thus displaced may now be found in Amazonia.

Rural Latin America exhibits classic uneven development: contrasts may be as great between adjacent rural areas as between city and countryside. Specialization both by individual farmers and by regions is promoted by changing markets, by new technologies and by state policies. The spatial division of labour – who does what, where and how – is reallocated. Each rural area has its own new role in supplying some combination of agricultural products and labour to the wider economy and in forming a market for goods from other areas. Many apparently very unsuccessful areas are important continuing sources of labour, for people as well as goods move in the rural world.

Agrarian systems

To comprehend a rural landscape, we need to understand the *agrarian system* which links the physical conditions for rural production to the social and economic (Fig. 8.2). At a given site, climate, relief and soils offer different opportunities under different technologies; but what is grown depends just as much on human demand, whether for consumption in some remote subsistence village or for sale to the cities or for export, and on human supply of inputs, from steel axes to fertilizers, tractors and pesticides. Similarly, the relations of production are as important to the landscape as the physical environment or the prevailing technology. Who does the producing, and who gets what is produced? How is the work organized?

The Colonial Period (Ch. 2) left a complicated pattern of land tenure across the continent, with large estates and community land as the commonest features. The three main changes before the Second World War were the abolition of slavery, the individualization of much community land into small, family holdings (so that the *latifundia-minifundia* system became extremely

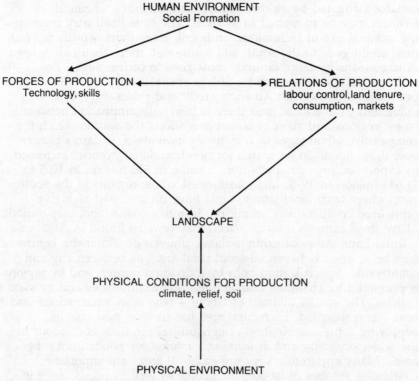

Fig. 8.2 Agrarian system

common) and the appearance of the modern plantation. In large areas of Latin America thirty years ago, most agricultural land was in large estates: haciendas, plantations and ranches.

Three main agrarian systems may be identified. First, the *plantations* in their most typical twentieth-century form were highly commercial: agricultural estates with a wage-labour work-force, producing specialized crops for an urban market, often overseas, using abundant capital and advanced technology, maximizing profits but at great risk from competition and market fluctuations. Banana plantations, made possible by refrigerated ships, perhaps fitted the model most closely, but certain similarities with the colonial slave-plantation for sugar or cacao are apparent. Our second system, the *haceinda*, is often contrasted to the plantation: it is an agricultural estate producing a range of crops for the local market (often mining or urban), using limited capital and traditional technology and minimizing risk. The work-force would work primarily for the exclusive use of small plots of land, not for cash; technology, transport and marketing would be highly labour-intensive; the hacienda could expand production when markets were good but could survive almost without inputs when they were bad. Early ranches were like haciendas in functioning almost without cash payments within the estates: in north-east Brazil or in nineteenth-century Argentina, workers might be paid in young stock and subsistence plots, under systems comparable to those of sheep haciendas in highland Peru. More recently, intensification of production on ranches by fencing and improved breeding of stock has gone with a change to wage labour. *Peasants*, our third system, were involved in both plantation and hacienda systems. Plantations had a permanent workforce but needed – and still need – much seasonal labour. Seasonal labour comes to banana plantations in Central America and Ecuador, to sugar in Peru, and to cotton in Bolivia from areas where farms are too small to support a family. The peasants who live on tiny farms, *minifundia*, and sub-family farms, *microfundia*, must find some work elsewhere for part of each year. In Peru it was estimated in 1971 that mountain peasants made up 57 per cent of the population, and received only 13 per cent of the national income.

During the last thirty years, a new agrarian system has entered the landscape: *agribusiness*. Agribusiness as an economic activity is defined as 'those individuals and organizations engaged in the production, transport, financing, regulation and marketing of the world's food and fibre supplies'. Giant corporations have now integrated these activities into a novel seed-to-consumer system. A traditional plantation may be foreign-owned, but simply produces, ships and sells its products on the wholesale market. Production may now be the least profitable part of the seed-to-consumer system. An agribusiness may even be less interested in the production of crops than in supplying inputs (seeds, fertilizer, machinery, pesticides) and in handling outputs (purchasing, shipping, processing, packing, retailing). Agribusiness can sell chemicals and aerial crop-dusters to plantations; it wants both haciendas and peasants to become much more regularly dependent upon inputs; it has persuaded peasants, particularly in Mexico and Central America, to give their land over to the production of crops on contract. In south-east Brazil, agribusiness has created a whole new crop landscape. It is difficult to find a farm in Latin America now not touched by agribusiness. It is agribusiness which has carried the Green Revolution across the continent to profitable areas.

Agrarian transformations

Agrarian systems throughout Latin America are changing from diversified, low technology, non-wage-labour, relatively labour-intensive systems powered by human and animal energy, to specialized, high-technology, capital-intensive systems powered by fossil fuels. The pattern is highly uneven, but everywhere there is a parallel change in the control of resources, in relations of production and in structures of social relationships; technical change is never independent of social change. Commonly there is a large increase in the role of wage labour; in many cases, all labour becomes wage labour. Models of change can usefully be applied to the types of agrarian systems already outlined: hacienda, plantation and peasant farming.

Models of agrarian transition

The owner of a *hacienda* who wishes to change from low investment and low yield to high investment and high yield has to change the relations of production – the way in which labour and perhaps land are controlled. Much of the hacienda land is being used by the labour force to produce food and rent; this must make way for specialization of land to the most profitable products and of labour to skilled tasks. In theory, two options are open: in Model One, the tenants can become wage labourers, while in Model Two part of the estate can be sold, to the tenants or others, and the proceeds used to capitalize the rest of the estate. Model One is really only possible if the hacienda owners have very strong political control, locally and nationally, for the tenants will resist losing their lands. In Peru, the hacienda survived into the 1960s just because the tenants were able to resist this change: they regarded their plots on the hacienda as a much more secure source of livelihood than joining the national labour market. The transformation of the whole estate, Model One, has been an important path of change in Colombia and much of Brazil, especially in São Paulo state; it may also be found in Central America, in coastal Peru before 1969 and in Ecuador and Venezuela.

Much more common has been the contraction of the large estate into a new, much more productive and often more profitable enterprise: Model Two. In Colombia, landowners have sold idle land which might have been taxed or confiscated, or even invaded by peasants. Elsewhere, intervention by the state by way of agrarian reform has often played the same role: the landowner has been allowed to retain the main buildings and an area of land up to a specified size, and compensation for the rest has financed the capitalization of the new enterprise. Less commonly, but significantly in Brazil, Colombia and Ecuador, urban profits have been used to buy and capitalize land, with the same result.

Plantations, the other variant of the large estate, may have started out as high-technology enterprises and simply continued on that path. Plantations with a longer history, such as sugar plantations in north-east Brazil, the old coffee fazendas of south-east Brazil and the henequen plantations of Yucatán have undergone changes (Model One or Model Two) in the same way as the haciendas.

How do *peasants* change to profitable, specialized agriculture? Can a small farm sector survive indefinitely alongside large estates, or will the small farms be swallowed up in the end, either by the large estates or by their more successful neighbours? This is an immensely important question for policy makers, since history and recent agrarian reform together have created very large numbers of tiny farms in Latin America. These very quickly divide into the more and the less successful, but the less successful have a remarkable ability to survive. (Functional dualism, one explanation for this survival, will be discussed below.) Peasants, under Model Three, continue at first to produce food for themselves but become more and more involved in the market. As they buy pesticides, fertilizer, the seeds of new varieties, even food, the time comes when they must sell produce in order to buy the inputs for the next crop; they cannot withdraw from the market when times are bad, and may even have to respond to lower prices by producing more. Specialization is the road to profit but not (Ch. 3) to security. The peasant is always in danger of being forced into borrowing at high interest; industrial inputs increase this risk. Even a small peasant needs then to be not just a skilled farmer but an entrepreneur, able to calculate costs and benefits and to relate to moneylenders, banks and government agencies. The urban-industrial sector now both buys farm output and supplies essential inputs and finance. Pearse (1980) has characterized the outcome as the 'talents effect'. 'To him that hath, it shall be given; from him that hath not, even that which he hath shall be taken away'. The peasant with more land, a larger family to work on it, more education, more contacts in the right places can gain from urban-industrial inputs. The landless may lose even their rented or share-cropped plots in a morass of debt, and debt costs many a peasant his land. A new development, especially in Central America and Mexico is production contracting: the direct introduction of new crops and techniques by agribusiness to peasants who grow a specified crop under specified conditions for sale only to the agribusiness, which may also supply all inputs and even credit and insurance. Are these still independent peasants, or have they become wage labourers with extended hours and responsibilities?

Many Latin American governments have not seen peasants as likely to achieve high output and incomes. We shall see agrarian reform policies which distribute land and create more small farms, but agrarian reform in Cuba, Peru and Nicaragua and on occasion in Mexico and Chile, has also meant that the state has maintained, expanded and even created large estates. The transition to these large state, collective and co-operative units constitutes Model Four. International agencies have also been in favour of co-operatives, which have been tried widely in Latin America but generally very unsuccessfully. Peasants are probably never a homogeneous group and are certainly not so under agrarian change: very quickly, some become successful entrepreneurs and others lose their land. Within a co-operative, interests differ, conflicts arise and almost always the better-off capture the benefits, sometimes even to the direct disadvantage of the poor; this has been particularly well demonstrated for the village of Matahuasi in the Peruvian Sierra. Model Four then becomes Model Three.

These four models of agrarian transition in modern Latin America may be briefly summarized:

Model One: Estate is converted directly to wage-labour, high-energy production (Lenin's 'junker road', called after the Prussian experience).

Model Two: Estate contracts in area and invests the proceeds of selling land to intensify production. Small farms may develop on the land sold. The state may force this process as agrarian reform.

Model Three: Small farms change from labour-intensive to capital-intensive small-scale farming. Less successful farmers become landless, or are forced to sell some or all of their labour for part of the year (Lenin's 'farmer road').

Model Four: The state maintains or creates large estates as state farms or on a collective or co-operative basis.

Interpretations

Two interpretations of Latin American economies which apply to rural change are modernization and functional dualism. Both point to the present contrast between the modern, high energy, capital-intensive agriculture and the poverty of smallholder agriculture but they give rise to different policy prescriptions.

Modernization theorists expect the modern sector to expand until it swallows up and replaces the traditional or peasant sector. In the process, it will exploit the peasant sector, but in the end the peasants will all become either successful farmers or wage labourers and the increase in production, the shortage of labour and high wages will eliminate poverty. Brazil subscribed to this view from 1964 to 1985 and concentrated on expanding production of profitable crops.

216

Functional dualism, however, considers that the family and sub-family farms are functional to the modern sector and essential to its well-being, and, indeed, that they exist for its convenience. They supply the seasonal and even permanent labour for modern farms and much temporary labour for the cities. Often the land is worked by the women, the very old and the very young. It is argued that workers whose families can feed themselves can be paid much less than if they had to feed their families: a part of the family livelihood is covered by family land. Peasant agrarian systems are thus said to be subsidizing the cities and the high technology farms by providing cheap labour. On sugar plantations in the Cauca Valley, Colombia, the majority of workers come from peasant farms, whether they be casual or 'permanent' (that is, on contract for a year or more). They are not true wage-labourers but part-time peasants, and are paid correspondingly little. The dualism between traditional and modern sectors is highly functional to the modern sector.

In agricultural production, functional dualists argue that peasants tend to produce staple foods for the urban poor; prices are kept low, again subsidizing wages. Modern farms produce the export, luxury and industrial crops and certain crops where massive production gains can be made, such as sugar and rice. This division in farm sizes, crops and technologies is recurrent; we have already seen it between hillside and flatter lands from Mexico to Bolivia. In Brazil, farms larger than 20 ha (50 acres) grow more than half of all the bananas, cacao, coffee, rice, sugar, corn, soya and wheat; smaller farms grow more than half of the manioc and beans. If Mexico is divided into irrigated and rain-fed areas, most peasants are found to live in the rain-fed zones which provide labour to the modern irrigated sector. Modernization theorists see all this as a transitional stage in modernization, but functional dualists see the nature of the traditional sector as contributing to the growth of the modern sector in a multiplicity of ways. Again, the spatial division of labour between agrarian systems is apparent.

Agrarian reform

Under the Alliance for Progress, from 1961, agrarian reform became a condition for United States aid to Latin America. Reform was to avert further Cuban-style revolutions; since production per hectare was substantially higher on small holdings than on large, an agrarian reform which redistributed land seemed to promise gains in both production and welfare. Mexico, Bolivia and Cuba had already had substantial reforms; almost all countries now passed agrarian reform laws, but few were implemented. Table 8.4 lists those reforms which affected more than a tenth of the farm population. Whatever the intention, these reforms have been significant in agrarian transitions; they have contributed to political stability and, in the long run, they have increased production. Some would argue that with the exception of Cuba (and possibly Nicaragua) the reforms have also

Table 8.4 Major agrarian reforms in Latin America

Country	Date*	Ceiling† (ha)	Impact‡	Character
Mexico	1917	100	1/2	Division of haciendas. About half of all farm land was reformed into communal tenure with individual rights. Model Two; some Model Four in the 1930s.
Guatemala	1952–54	90		Idle land and rented land transferred from large estates to small farmers and co-operatives. Reform reversed after 1954 coup.
Bolivia	1953–70	10	2/5	Division of haciendas. Much of the cultivated land was transferred to tenants. Model Two, Model Three.
Cuba	1959–63	67		Initial expropriation of idle land, and land transferred to tenants; all large farms later converted to state farms. Model Four.
Venezuela	1959–70		1/6	Some division of haciendas, much settlement of newly irrigated land. Model Two, Model Three.
Chile	1967–73	80	1/5	Division of haciendas. Co-operatives and after 1970, collectives encouraged. Model Two, Model Four. Reform reversed after 1973 coup. Model One, Model Three.
Peru	1968–76	35	1/3	Division of haciendas, expropriation of plantations. Much promotion of co-operatives. Model Three, Model Four.
Ecuador	1973			Some division of haciendas, land to tenants especially on the coast; much sale of land. Model Two, Model Three.

Table 8.4 contd.

Country	Date*	Ceiling† (ha)	Impact‡	Character
Nicaragua	1979	350		State farms on the lands of the ex-dictator, expropriation of idle land, land transferred to tenants, much promotion of co-operatives. Model Two, Model Four.
El Salvador	1980	245	2/5	Division of very large estates, some land to small-scale tenants. Model Two, Model Three.

* Date of the main reform law and the end of the reform
† Largest size of farm allowed on the best land
‡ Estimate of the fraction of the total agricultural population affected

promoted functional dualism and with it the exploitation and immiseration of many. Agrarian reform has been interpreted as a victory for the urban-industrial sector over the landlords, as it has altered the structure of political power and promoted agrarian transition. In Cuba, Mexico and Bolivia, landlords did indeed lose wealth and power, but in Mexico and Bolivia a new, very wealthy agrarian élite has appeared. In other countries, notably Ecuador, reform has proved a convenient route for hacienda owners to Model Two transformations. Like technical change, agrarian reform proves not to be an unambiguous benefit to the small farmer and the landless. In real agrarian systems, both technical change and agrarian reform promote increases in overall production, but neither necessarily assists the livelihoods of the poorest.

Despite the early date and the scale of the reform, Mexico has more desperately poor peasants now than before the revolution of 1911–17. The agrarian reform, by pacifying some of the landless, is said to have made possible Mexico's great urban and industrial revolutions, for agriculture earned essential foreign exchange, supplied cheap food and kept wages low. A significant innovation was that beneficiaries of the reform were not able to sell the land. The land belongs to the community (*ejido*); farmers hold land which may be passed on to heirs, but not sold; if the land is not used, it reverts to the community. This is intended to prevent the talents effect of Model Three, but a series of new developments poses serious problems. Private owners of land can use it as collateral for credit to buy the new inputs. Agrarian reform beneficiaries cannot, since the land is not theirs to sell; many now rent out their land instead – in some states of the north-west, 90 per cent of reformed land is so rented. A new kind of entrepreneur has appeared, with capital for the inputs but without land; many now control large estates. The surviving cores of many old haciendas also prosper. Some evaded reform: many groups of farms under the legal size limit are held by straw men but actually worked as one estate. Production contracting is widely used by agribusiness, since foreigners may not hold land in Mexico. The 1980s have been compared with 1910 in Mexico: foreign capital and large national concerns are active in rural areas, landlessness is rising and malnutrition is widespread though large gains have been achieved in production.

Peru has seen the most extensive agrarian reform in South America. All large estates were expropriated and no land was left to the estate owner; some 10 million ha (25 million acres) changed hands. The coastal plantations became co-operatives owned by their permanent workers; haciendas in the Andes were merged into great rural co-operative enterprises. Peasants with sub-family farms and landless peasants without permanent employment on an estate were excluded from the reforms and now work for the minority who did benefit, on terms as adverse as before. There are also serious management difficulties and production has not risen as the urban-industrial élite would wish. This was explicitly a reform for greater productivity and for better welfare; so far it has achieved little of either.

Cuba attempted a complete transformation to socialist organization of rural production, Model Four, on the traditional socialist conviction that only large agrarian units are efficient (a tradition now challenged by China). All large farms were taken over by the state; peasants were encouraged to form unions and co-operatives, and indeed to disappear as independent producers. Only since 1980 have direct sales of food by peasants been legal; the peasant and co-operative sectors are now being given much more autonomy and encouragement, and in a sense, Model Three is alive and well in centrally-planned Cuba. Cuban agriculture has not met Cuban aspirations in growth, but there has been great success in raising rural living standards, eradicating acute poverty and achieving a large degree of equity not only within rural areas but between rural and urban areas.

In Mexico, reform promoted agrarian transition; in Peru, it seems so far to have failed to do so; in Cuba, gains have been more impressive in welfare than in production. Brazil and Colombia have achieved dramatic increases in production and significant agrarian transitions without more than token concessions to reform. Some Latin American states have clearly seen agrarian reform as essential to economic development; some have made only trivial use of it. Despite this contrast in state strategies, all except Cuba still have large numbers of tiny farms yielding a meagre living and large numbers of landless workers without even secure employment, let alone a decent wage. A major transition is nevertheless over, for there is now a dominance of agrarian systems based on specialization and wage labour. Debt peonage and service tenancy (Ch. 2) are legally features of the past; rural labour in Latin America is now very effectively controlled by the market system, often in conjunction with a repressive state which restrains strikes and land invasions. In Guatemala, peasants used to be compelled to come from the mountains to work on the banana plantations; now they must come for the wage, for their tiny farms are too small and eroded to support their families. In the agrarian transitions the role of extra-economic coercion has been greatly reduced, but the welfare gains may be small.

Colonization

The development of agriculture on previously non-agricultural land has contributed greatly to growth in production since 1960 (Figure 8.3). Large areas have been affected, but relatively few people: official colonization schemes absorbed perhaps 2 per cent of the increase in rural population from 1960 to 1980 and spontaneous colonization perhaps another 20 per cent. Far larger numbers went to the towns (Ch. 9). Colonization is perceived by politicians and agencies as an escape, a new opportunity to create new rural worlds, but, in colonization areas, latifundia-minifundia patterns reappear, with all the associated inequalities in access to credit, inputs and expertise.

Some 54 per cent of the world's remaining tropical moist forests are in Latin America: it is clearing on this frontier that has raised international alarm. There are unfounded fears for the world supply of oxygen, misgivings about global climates (probably threatened more seriously by the burning of fossil fuels) and alarms about the loss of species, as some 10 per cent of the world's living species are found in Amazonia. These species, however, survived through the Ice Ages when the forests contracted to a fraction of their present area: similar refuges could be designed to protect them again. Time would be essential for this, and only the world recession and the relatively high cost of oil have slowed the march of this frontier. The immediate danger is a local change in rainfall. Half the rain that falls in Amazonia is recycled by the forest; it does not reach the sea. The water that leaves the basin at the mouth of the Amazon is equivalent to an average rainfall of 800 mm (31 in) – yet actual rainfall varies from 1,500 to 4,000 mm (59 to 156 in); the difference is evapotranspired by the forest. In Latin America, local and continental effects from extensive tropical forest clearance are a real possibility.

The soils beneath the forests of the tropical Americas are extremely varied. Generally, physical conditions for agriculture are good but chemical conditions are poor, calling for sophisticated management with traditional methods (Ch. 3) or chemical fertilizers. Over much of Amazonia, plant nutrients are extremely scarce, so that the forest traps the nutrients from the rain before the water

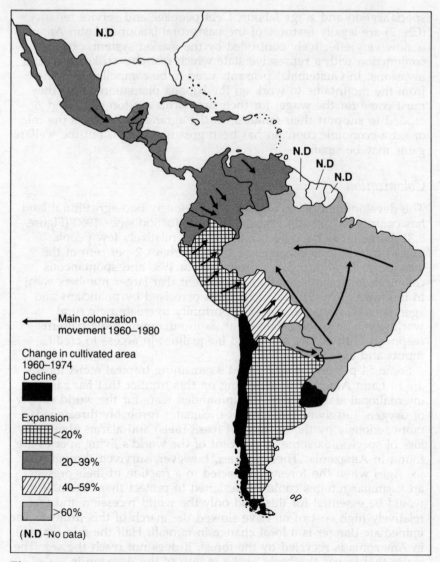

N.D

N.D

N.D

N.D

← Main colonization
movement 1960–1980

Change in cultivated area
1960–1974
Decline

Expansion
|||| <20%

▓▓ 20–39%

/// 40–59%

░ >60%

(**N.D** –No Data)

Fig. 8.3 Colonization

escapes into the soil: the rain is richer in nutrients than are the
rivers. It used to be thought that the 75 per cent of Amazonian soils
which are acid and of low fertility would always decline to very low
productivity and perhaps harden to laterite if cleared of forest. In
fact, only about 4 per cent of the subsoil is plinthite, which will
indeed harden irreversibly if exposed and dried. Experimental work
has shown that agroforestry is practicable, that rational pasture
management will maintain fertility and that even small crop farms
can be viable. At Yurimaguas, in Peru, since 1971 soil scientists from

North Carolina State University have advised colonists on land that would normally have been abandoned after a few years. Close monitoring of the soil has enabled the colonists to purchase and apply appropriate fertilizer and to maintain profitable production of crops for market, even with simple technologies. These remain, alas, undeveloped possibilities: agroforestry, pasture management and soil monitoring have reached only a tiny fraction of the cleared area, and have received a miniscule portion of the sum invested. The real recurrent pattern from Mexico to the Mato Grosso is that colonists clear the forest for crops, fertility declines, weeds invade the area, credit is unavailable and the colonists fail or are driven off by force in favour of extensive cattle ranching. This degrades the ecosystem into low productivity grassland; the scarce nutrients are diverted into cattle and transported away as meat; employment is minimal. Only the speculative value of land permits this to happen. Absentee landowners extract the profits by extensive land use; employment is low and welfare poor. Until 1979, Brazil subsidized these practices in the interests of developing Amazonia: between 1966 and 1978 the regional development agency authorized $950 million in tax rebates to 358 cattle ranches; these had acquired 800,000 sq. km (308,800 sq. miles) of land, or ten times the area of Austria (Table 8.5).

Table 8.5 Land tenure in Eastern Amazonia

Farm sizes (ha)	Agricultural areas		Livestock areas	
	% of land	% of farms	% of land	% of farms
< 100	62.7	94.0	6	69.2
100–999	24.7	5.2	9	24.1
1,000–9,999	11.0	0.7	29	5.2
10,000	1.6	0.1	56	1.5

Source: Hecht (1983)

With time, refuges for species could be identified and defended and employment in the already settled areas: it is a political choice to use the forest as a safety-valve. Most colonization is spontaneous, and securing real gains in human welfare, but almost everywhere there is pressure for forest clearance. The state builds and paves roads, investors seek pasture and minerals, the colonists are pressed against the forest edge. Most countries could increase production and employment in the already settled areas: it is a political choice to use the forest as a safety-valve. Most colonization is spontaneous, but all is highly commercialized and depends on roads to the market; where there are no roads, there are no colonists. The small farmer who fells the forest is called a threat, but he is more of a

223

Plate 12 The frontier: a small farm created from the forest on public land, on a tributary of the Magdalena River, Colombia. Farm and forest have since been replaced by ranching. (Photo: J. Townsend)

Plate 13 The frontier: the end of the road and the beginning of the mule trail: a Colombian village takes shape as the forest falls. (Photo: J. Townsend)

tool, in position because he had been displaced elsewhere and because roads have been built. The responsibility lies with the state and with modern sector farming. The details vary in social, economic, legal and physical conditions, but colonization tends rather to extend the rural slum than to create the stable family farm beloved of so many government agencies. The exciting resource frontier becomes a mere misused periphery. Modernization theorists seek more efficient forms of colonization, hoping to reduce the failure and appalling suffering of the pioneers. Functional dualists see that very failure as part of the system, clearing land cheaply for the cattle which are sold in the more profitable markets in the cities and overseas, and producing cheap staple foods in the process from the ashes of the forest.

The State

Agricultural research, high energy technology, agricultural extension, credit, agrarian reform, colonization, road building, tax and price policies: all these influences in the rural world must be examined together for specific times and places if we are to understand landscape and rural living. In each sphere of influence the state must play a role; often it decides between them. With the exception of Cuba every Latin American country which had a major agrarian reform has subsequently made its main rural investments outside the original reformed area. Mexico created a whole new irrigated landscape in the arid north-west, far from the congested highlands, now she too has turned to the tropical forests, for cattle. In Bolivia in 1953, almost all the land to be reformed and 90 per cent of the population were in the highlands and valleys. Roads and growth have since gone to the eastern lowlands: cattle ranches around Santa Cruz have become cotton estates (Model One) and colonists to whom land was given under the reform have carved rice and sugar areas out of the forest around; when the land is used up, cattle take over. In Venezuela, the reform consisted mainly of settling highland peasants on lowland irrigation schemes; in Colombia, more cheaply, of giving title to colonists who had settled public lands. In Peru, the state is now selling vast tracts of the eastern forests – tracts often occupied by Indian tribes, colonists or both, to transnational companies. Brazil has the vast Grande Carajas scheme in eastern Amazonia, offering heavy tax concessions and subsidized power and transport to mining, industrial, forestry and agricultural investors. No small colonists in Brazil have had comparable support. Programmes to run cars on alcohol made from sugar cane threaten to take up a fifth of the cultivated land in the north-east and south-east, forcing up food prices: again, a state decision.

Everywhere the state intervenes constantly in prices, both for domestic consumption and by subsidizing or taxing exports. The rise in wheat yields in Mexico from 750 kg/ha (2 tons/acre) in the 1940s to 4,200 in the 1970s came not only from new seeds but from favourable prices: fertilizers were kept cheap and wheat prices high. Surplus wheat was exported with a subsidy to bring it down to the world market price. When wheat prices were allowed to fall, the wheat area contracted and Mexico returned to importing wheat: consumers gained, producers lost. The state intervenes between groups not only in prices but in agricultural research. Hybrid maize and new varieties of rice in Colombia worked to the advantage of large farmers in Valle del Cauca and to the disadvantage of small farmers elsewhere. This may seem a questionable form of rural change, but cheaper staple foods benefit the poorer income groups, particularly in urban areas. Food prices can often be a major political issue and usually urban people can bring more pressure to bear on the state than rural people. Latin American governments consistently keep the price of staple foods low, creating a short-term gain for the poor, but also helping to keep wages low and providing little incentive for increased production. Small farmers in Latin America commonly regard the state as an enemy. Certainly, state decisions on public investment, legislation, prices and taxes penetrate every aspect of rural life and rural change.

Conclusion

Rural change in Latin America is not a temporary phenomenon: these are shifts with long-term geographical implications. They reflect national processes of socio-economic transformation and transnationalization which have specific local outcomes. They may themselves account for long-term future changes in the national and international scene. The particular agricultural products exported from before the First World War, in Latin America's heyday as a primary exporter, were the direct cause of much of the human geography of the region well into the 1960s. Temperate agricultural commodities from the extensive use of land in Argentina and Uruguay called for a good transport net; the products competed with those of other temperate regions with a higher standard of living; labour shortage required immigration; growth in production and relatively high wages promoted the domestic market and industrialization; industrial and urban changes followed which sufficed to keep these countries among the richest per head in Latin America. Tropical agricultural exports, on the other hand, entered world trade in competition with those of colonies and of the slave-holding areas of the United States; prices and wages remained low and a complex infrastructure was rarely required; conditions for industrialization were not created. The exceptions were coffee around São Paulo and to a much lower degree in Colombia, where local conditions promoted infrastructure and, for very different reasons, a domestic market, in both cases these together were conducive to industrialization and urbanization. Historically, the specific incorporation of certain rural areas of Latin America into the world economy at a given time explained the urban and industrial development of whole nations for decades to come. Rural change is important not only to rural people.

If world recovery takes place, different rural areas will find that their national roles have shifted, and that the ways in which they are linked into the international economy are altered, as well as their vulnerability to it. Now that rural areas have lost so much relative importance, their present role may not be as fundamental to the future of their countries in the world economy as was the case in 1900, but their roles in the spatial division of labour and the social and economic structures of their countries will still be important and the management of these roles may be momentous at the national, regional and individual level.

Further reading

Altimir O (1982) *The Extent of Poverty in Latin America*. World Bank Staff Working Paper No. 522.

Bronstein A (1982) *The Triple Struggle: Latin American Peasant Women*. WOW Campaigns Ltd, London. (A vivid evocation of the lives of half the rural population in the Spanish-speaking republics.)

Foweraker J (1981) *The Struggle for Land: a Political Economy of the Pioneer Frontier in Brazil from 1930 to the Present Day*. CUP, Cambridge, New York, Melbourne. (Excellent but heavy going.)

Goodman D and **Redclift M** (1981) *From Peasant to Proletarian: Capitalist Development and Agrarian Transition*. Blackwell, Oxford. (Fine theoretical discussion of agrarian change in Latin America with excellent analyses of Mexico and Brazil.)

Hecht S (1983) 'Cattle ranching in the eastern Amazon: environmental and social implications' in **Moran E** (ed.), *The Dilemma of Amazonian Development*. Westview Press, Boulder, Colorado.

Kutcher G P and **Scandizzo P L** (1981) *The Agricultural Economy of North-east Brazil*. A World Bank research publication. Johns Hopkins University Press, Baltimore and London. (A quantified argument in favour of land reform.)

Lindqvist S (1974, translated 1979) *Land and Power in South America*. Penguin, Harmondsworth, New York, Ringwood, Ontario, Auckland. (Fine journalism with much individual illustration.)

Moran E (ed.) (1981) *Developing the Amazon*. Westview Press, Boulder, Colorado. (Very useful up-to-date interdisciplinary collection.)

Pearse A (1980) *Seeds of Plenty, Seeds of Want: Social and Economic Implications of the Green Revolution*. Clarendon Press, Oxford.

Preston D A (ed.) (1980) *Environment, Society and Rural Change in Latin America*. John Wiley and Sons, Chichester, New York, Brisbane, Toronto. (A useful collection of papers on migration, agrarian reform, colonization and agribusiness.)

CHAPTER 9

Population mobility and the creation of new landscapes

David Preston

Latin America, like Anglo-America, is distinctive among other world areas in that most of its population is descended from people who migrated within the last 450 years from Europe, Africa and eastern Asia. The style of livelihood that the newcomers adopted reflected the nation, region or cultural group from which they came as well as their means of making a living. Some immigrants came in large groups and formed cultural islands in a sea of Latin Americans retaining the language and values of their country of origin. Nevertheless, they all eventually became citizens of Argentina or Mexico or of whatever their new country: their children invariably spoke Spanish or Portuguese as well as their mother tongue. Within a couple of generations a high degree of assimilation had taken place. In spite of this, the livelihood of the children and grandchildren of the migrants might well still depend on ice cream, ranching or banking just as it had in the time of their immigrant grandparents. Since immigrants tended to intermarry in the first generation, their kinsfolk would largely comprise people from the same area and their circle of friends might still be composed largely of the descendants of other settlers also from the same area.

These events mean that a major feature of many Latin American areas – in cities, towns and countrysides – is the different cultural characteristics of the human population which are reflected in the ordering of the settlement, the style of house and the type of business. It is therefore necessary to understand the social, economic and geographical origins of settlers in order to start to comprehend present-day relationships between the population, its ways of making a living and the city or rural environment. Many areas such as São Paulo state experienced waves of immigrants from different countries at different times. The composition of the migrant groups in the past needs to be known in order to explain the circumstances in which certain groups of migrants came and the ways in which they established themselves.

A second aspect of population mobility which is important to an understanding of socio-economic processes concerns the movement

229

of people within large geographical regions of Latin America; from long-settled rural areas to new lands or to towns and cities; from towns to cities and from various urban environments to new land. Such movements are part of the strategies that people adopt to enable them to live at a satisfactory level and the changes in such spatial and socio-economic patterns through time, even in a decade, reflect and reveal important changes in the human and physical landscape. If people are influenced in various ways by the places in which they live and work, so too is the land and the built environment influenced by the use to which people put it.

Migration in the past

The course of conquest

With the start of the conquest of Latin America by groups of Spaniards a new cultural element was introduced into the human population. Except for the Inca empire no large area of Latin America was a large coherent, organized political unit. The Spaniards, and those with them, represented a single imperial nation and, although there were many conflicts between Spaniards, including periods of civil war, they were from a broadly similar cultural background; more than that, they were established in the continent by virtue of being conquerors and were a living reminder to the native population of their subordinate status.

The Spaniards who went to the New World were from a broad cross-section of Spanish society. No social class was without its representation: a wide range of jobs were done by Spaniards even though lowly labouring jobs were carried out by African or local Indian slaves. During the period of conquest, in the first half of the sixteenth century, the paucity of Spaniards and the considerable areas of land to which the conquerors could obtain some form of access meant that individuals of relatively humble social origins might become *encomenderos* with a right to receive tribute from subject Indians but without direct access to their lands. They passed into the landed upper class as Indian numbers declined with a degree of ease that was not available to similar and equally valiant persons coming to the New World one or two generations later.

The lengthy period of Spanish rule spanning more than two centuries was one during which Spanish migrants and their descendants controlled the economy of the continent and the society of the towns and cities. In the countryside, by contrast, Spaniards were scarce and a largely Indian population existed, perhaps reorganized by the Spanish, but with their own social structure apart from that of the conquerors. The Indian lords who had not been killed retained some of their status and were recognized as Indian leaders but enjoyed only the equivocal status of being the highest among the lowest.

The Spanish migrants to the New World included nobles, soldiers, craftsmen and merchants as well as clergy and bureaucrats and they effectively recreated an American version of Iberian society. The Spanish came from Castile and Andalucia but also from other parts of Spain. By the mid-sixteenth century Andalucia became the most important source of migrants as revealed by the registers of emigrants on ships leaving for the New World. It is asserted that they settled particularly in the coastal towns and inevitably their distinctive pattern of speaking influenced lowland speech; but others came too who were not among the 54,881 official migrants from Spain to the New World between 1492 and 1600. They included not only other Europeans working with the Spanish but also the necessary imported slave labour force. In the period following the conquest many slaves in Peru were Indians brought from Central America, but alongside these slaves were blacks imported from Africa. The African slave trade brought some 11 million people to the New World until the ending of slave trade and slavery during the nineteenth century. The black migrants provided an urban labour force throughout the empire working alongside Indians and mestizos and often with highly-prized specific skills. Some West African slaves were skilled metal workers and had a team of workers under their direction making high quality goods from whose sale their owner derived profit. In the New World black people were particularly used for agricultural labour where no local labour was to be found. Thus in the Portuguese colonies on the Atlantic coast which eventually developed a plantation economy based on sugar cane the labour force was largely black as is the population there today.

By 1800 there had been a substantial redistribution of the population by comparison with the situation in 1500, as has already been described in Chapter 2. Spanish and Portuguese had settled almost all the areas where temperate and sub-tropical agriculture was possible as well as some lowland coastal areas. The Europeans were still concentrated in the urban areas and, although the Indian population was considerably smaller than that of 1500, in densely populated areas where Indians did not predominate, it was a mixed-blood people rather than Iberians who predominated, although the precise cultural origin of the people was usually less important than their social position. Those coastal lowlands that were densely populated were predominantly peopled by blacks or descendants of African slaves although here, too, white Hispanic people were located predominately in urban centres and in middle and upper social strata.

Nineteenth-century immigration

By the time that Independence movements were growing in importance in the early nineteenth century, the frontiers of settlement in eastern and southern Latin America were still relatively close to the coast. In Argentina Indians presented a serious drawback to settlement in the Pampas until the last part of the century, even though there were few new settlers. In Brazil mineral discoveries (gold and diamonds) had led to settlement in the uplands of Minas Gerais by 1700 and a shift of centre of gravity both westwards and southwards and a wave of Portuguese immigrants in the last half of the eighteenth century was one of the causes of a ten-fold population increase during the eighteenth century. Most of the migrants were men and widespread miscegenation resulted rather than the creation of a cultural enclave. But in 1800 São Paulo was still a frontier town belonging to the *bandeirantes*, armed bands of wandering slavers and traders, even though coffee was being planted in the Paraiba valley to the north using slave labour for export through Rio. The southern part of Brazil, like the west, was sparsely populated.

During the nineteenth century, the most important changes in the population of Latin America were not associated so much with the independence from Spain and Portugal but with the wave of European immigration that peopled some of the emptier areas in the southern cone (Chile, Argentina, Uruguay and southern Brazil); this provided a commercial and social stimulus as foreigners began to farm parts of the lowlands and to develop commerce, transportation and mining in other areas. Spain's commercial hegemony had long since waned and Great Britain, France and Holland became major traders for the continent as they sought markets for the goods and machines that the Industrial Revolution produced. The isolated areas of the empires – the south in Brazil and the River Plate in Spanish America – had long suffered neglect and restrictions on free trade and the severing of these controls increased the range of contacts that such areas made with the northern hemisphere. While the development of coffee and bananas as export crops in Central America in the second half of the nineteenth century was the result primarily of foreign investment and immigration, the latter was small-scale since the labour was locally available: Indian lands were being appropriated by non-Indian townsfolk forcing the Indians to seek at least seasonal employment elsewhere. Investment was primarily from the USA although some was from Britain, Spain and Germany.

While the abolition of slavery was also a powerful factor stimulating the development of new social relations with production, only in a few areas was the labour shortage solved by importing indentured labour. In Peru, for instance, in the middle of the nineteenth century, Chinese and later Japanese were brought in large numbers to provide labour for cotton and sugar plantations on the coast (Faron 1967). Chinese were likewise imported into Mexico, Cuba, British Caribbean islands and Guyana. They came from humble backgrounds but, as the size of the Chinese (and Japanese) community grew, other Chinese migrants arrived with greater economic resources in order to profit from trade with their home community. In Cuba the existence of empty areas with a good potential for farming was attractive and, even before Independence, Spanish-speaking migrants came there from Hispaniola (following a black rebellion) and Louisiana (when it was transferred to France in 1803). Migrants from France founded Cienfuegos in 1819 and many refugees from the civil wars attendant on Independence found refuge in Cuba. In addition, large numbers of Africans were imported into Cuba until the late 1860s despite growing efforts to end the slave trade.

The southern cone saw the most important transformation as a consequence of an increasing volume of immigration during the nineteenth century but this swelled to a flood during the last quarter of the century which continued undiminished until the start of the First World War in 1914. This phase of immigration was distinctive because it effectively settled a large temperate part of the continent with a predominantly European population. This settlement was closely related to the creation of new means of communication between Europe and the New World – steamships – and overland – mainly railways – in Argentina especially, with the use of foreign, mainly British, capital. In the southern part of Brazil colonization by Europeans was much less tied to railway or road building although in the coffee zone of São Paulo railways were just as important as in Argentina and eventually became major reasons for intensification of beef production beyond the Parana river.

Immigration into São Paulo state was closely associated with coffee production and in the 1880s Italians replaced Portuguese as the most numerous group (62 per cent) many of whom were subsidized migrants destined for the coffee zone. Most came from northern Italy (69 per cent in 1876–1900) at a time when southern Italians were predominantly migrating to the USA (Merrick and Graham 1979) and São Paulo planter organizations actively sought migrants from northern Italy. Many of the Italian migrants were whole households and the access to land, even as tenants, with the opportunity for various members of the household to earn wages at least seasonally, allowed many to become landowners or to enter business in the cities with some capital resources. By 1920 immigrants owned 27 per cent of all rural properties (and 19 per cent of the total property value) in the state of São Paulo; by 1932 immigrants owned 40 per cent of all the coffee trees in the state and the Italians alone owned 22 per cent (Merrick and Graham 1979). Further south in Brazil, immigration was rather different; it was smaller-scale and from a wider range of sources with a proportion of the immigration being spontaneous although state governments and private colonization companies were also important organizers or stimulants to foreign immigration. German settlement predominated at first and São Leopoldo, the first European colony, was founded in 1824 just north of Porto Alegre in the southernmost state of Rio Grande do Sul. After 1870 Italian colonies were established and in the last quarter of the century German, Italian and Brazilian colonies had settled a good deal of the forested tablelands in the eastern part of Rio Grande do Sul. In Santa Catarina, further north, another important German colony was established near the coast and by the 1890s settlers were establishing themselves further inland. During the same period the newly-created state of Parana was being settled by Germans, Italians, Poles and Ukranians stimulated by the construction of railways linking the newly cleared lands with the coast and with São Paulo.

The colonies were small scale and aimed at establishing family farms rather than large commercially oriented estates. Although some colonies were located where there was access to a market, others were isolated. The European settlers found the development of a distinctive style of farming difficult and many adopted a Brazilian style of subsistence farming and had little contact with the rest of the state and the nation. A negro worker on a German farm was reported to have learnt German and forgotten Portuguese so little contact did he have with non-German speakers. The foreign colonies were not sufficiently successful to allow the colonists to make a good living. As the years passed the succeeding generations moved further west to settle new land rather than subdivide an already small farm. The villages and towns acquired a distinctive appearance reflecting the cultural origins of the inhabitants but the chief role of the colonists from whichever country they came was to settle new land and to incorporate empty areas into the Brazilian nation.

In the river Plate countries independence early in the nineteenth century similarly created an awareness of the need to populate the country. The confirmed threat to farming presented by indians in the areas west of Buenos Aires discouraged settlement here until after their extermination in the Conquest of the Desert (1879–83) but in Santa Fé and Córdoba provinces immigration was encouraged to settle the land west of Rosario. Colonies were established along the Rosario – Córdoba railway by the Central Argentina Land Company in the 1870s and private settlement schemes also existed. New settlers moved the frontier northwards in the 1880s into Entre Riós but further settlement on a large scale did not take place until after the First World War when *quebracho* trees were exploited and settlers came from Paraguay in the north and from German and Slav areas of Europe.

The immigrants were not all farmers. They came from every part of Europe and included the nobility and the poor. Argentina, which had a population of 1,200,000 in 1860, received 2,500,000 immigrants in the next fifty years and by 1910 three out of every four adults in Buenos Aires was European-born. From southern Italy came the *golondrinas* (swallows) for the wheat harvest, single men who returned home after the end of the southern hemisphere harvest to work their own fields. Many migrants remained, some stayed in the rural areas of the Pampas attracted by tenancies which offered short-term security and the possibility of saving money.

Others stayed in the cities, in particular Buenos Aires. Eighty per cent of the immigrants were Italian and while southern Italians were predominantly attracted to rural areas the northerners stayed largely in urban centres.

In Uruguay, likewise, immigration facilitated the growth of the farm (and national) economy but the proportionate importance of immigrants was less than in Argentina which in 1914 had 30 per cent of its population born overseas compared with 17 per cent in Uruguay (in 1908) and the foreign-born population (mainly Italian and Spanish) tended to settle close to the river Plate and the river Uruguay (Crossley, 1983).

Twentieth-century immigration

The great tide of European immigration to Latin America that peopled the southern countries grew in the last quarter of the nineteenth century. In the twentieth century the same processes continued as land further from the main cities and agricultural regions was settled. In Brazil settlers reached the Mato Grosso and the margins of the Amazon basin and the settlement of the grasslands began in earnest. In Argentina settlement extended further north towards Paraguay in the Chaco, Misiones and Formosa where German-speaking migrants from eastern Europe came in the 1920s and 1930s. Irrigation works in northern Patagonia in the Río Negro valley and the arrival of the railway permitted more settlement and the extension of intensive agriculture further north in the Mendoza area created further opportunities in the early part of the century for migrants, many of whom were from Europe.

Elsewhere in Latin America small numbers of immigrants were establishing themselves as the cities grew and commercially oriented farming, sometimes organized by foreign companies (such as Grace in Peru, or the United Fruit Company in Central America), in lowland areas became more important. In some cases particular groups of immigrants such as the Germans, became associated with a single crop, such as coffee in Nicaragua or sugar in Cuba but usually their total contribution to production was small even if they produced a disproportionate amount of the best quality goods.

An important, yet seldom discussed, characteristic of twentieth-century immigration to Latin America is the diversity of migrant origins and small numbers of settlers from a particular cultural group have often made a very distinctive contribution to regional and national life. The most striking of these groups includes the various German-speaking migrants of the twentieth century. This includes Jews expelled by Hitler Germany, Sudeten Germans and Nazis and Nazi collaborators fleeing the consequences of the end of the Second World War. Many were relatively wealthy business people and skilled craftspeople and they found it possible to insert themselves into Latin American society and to achieve economic success. In the 1960s, but to a lesser extent in the 1980s, many of the better hotels, photographic businesses, and money exchanges were owned and directly run by German-speaking people. Thus, when seeking a good small hotel in the Bolivian Yungas, a photographer selling good-quality postcards or a source of dry Spanish sherry, the chances were that you would deal with a German immigrant.

A second small but important group of migrants similarly encountered throughout Latin America are the *turcos*. The term covers Levantine people many of whom came from Lebanon and elsewhere in the eastern Mediterranean after the First World War. Like the Germans the Levantines were victims of political upheaval and comprised middle-class families with the capital to get to Latin America and to start businesses once arrived. Almost exclusively urban, they established shops, typically selling a wide variety of products appealing to a range of customers but others became established in industry and in financial enterprises.

Each of these groups shares the common characteristic that their limited numbers and common cultural and geographical origin made collaboration between them straightforward and the nature of their role in local economies made it easy and advantageous for imports, currency, warehousing and marketing to be handled by others from the same group. Social gatherings were and still are common and the Yellow Pages in any Latin American city will list a series of social clubs that act as meeting places for people descended from immigrants from a particular area. This same phenomenon is, of course, characteristic of all migrant groups in the northern hemisphere as well as the south but it does enable small but important immigrant groups to be identified in every Latin American country.

Oriental migrants are highly visible in Latin America even though they are not numerous. Asians were brought to Latin America since the decline of the economic role of African slaves in the nineteenth century. Indentured labourers from south and east Asia were brought to areas of plantations to fill the gap left by the departing slaves. In parts of the Caribbean and the Guyanas the numbers of indentured southern Asians in territories with only a small total population meant that they formed in time a substantial proportion of the population and remain today as a sizeable and visually identifiable group in such areas. Asians are a powerful political force in both Trinidad and Guyana. Elsewhere in Latin America east Asians are the most commonly identifiable Asian group. Many came during the second half of the nineteenth century, both from Japan and China, and, although many came as contracted agricultural labourers and a proportion returned, others acquired the means to improve their status, to work as sharecroppers, purchase small-holdings but most commonly to return to the towns and to start small businesses without any link with their farming experience. The experience of the Japanese in the Chancay valley north of Lima differed from that of the Chinese. They became deeply involved with the spread of cotton growing, as sharecroppers, managers and as organizers of commercial enterprises from the first decades of the twentieth century. Anti-Japanese feeling during the Second World War provided a convenient excuse, as it did elsewhere in the Americas, for expropriation of many Japanese assets and for a decline in their importance in commerce and agriculture.

Elsewhere in Latin America, Asians have been important elements in the population that has settled new lands, especially in Brazil and Bolivia. The numbers have always been relatively small but groups of east Asians, both Chinese and Japanese, have often been seen as relatively more successful than Latin American migrants. There is only limited evidence to support this conclusion but the cohesive and well-organized nature of such culturally alien groups often gives them advantages over individualistic Latin American colonists receiving only limited guidance from government agencies. The most distinctive long-term contribution of these descendants of international migrants is the businesses which they now carry on, the most visible of which is catering. 'Chinese' restaurants are a feature of the urban scene in many parts of Latin America where Asians came in search of a better living than at home, but in other sectors of the business world too Chinese and Japanese have local and sometimes national importance.

The underpopulated areas of Latin America have long been a haven for groups suffering persecution in their homeland because of their beliefs. The best known group is the Mennonite religious group who migrated from eastern Europe to Canada, and some then moved to Mexico and to Paraguay, as well as to Bolivia, Belize and elsewhere. Seeking freedom to worship freely and strongly independent they formed successful pioneer groups and are now well-established providers of dairy produce (Belize), cotton or beef (Paraguay). The Welsh too sought freedom, both political and spiritual, in their settlement of part of Argentinian Patagonia in the late nineteenth century.

Movement between countries in the Americas

The arbitrary nature of many Latin American frontiers ensures that there is a great deal of movement of population across international borders but the similarity of the cultural and ethnic composition of people either side of the border means that the movements are a reflection not of the nature of the people but rather of economic and political differences whose natures change from time to time. Where major cultural divisions may seem to exist such as between Latin and Anglo-America (Mexico and the USA) such differences are much less remarkable on the ground. Much of the southern border regions of the south-west of the USA were formerly part of Mexico and the population of the adjacent areas of the USA is Latin American in origin.

Two broad categories of movement can be identified. In the first place it is a basic principle of migration that people move in response to their perception of opportunities for increasing their well-being and that of their households. Thus it is to be expected that one current of migration is towards countries and regions within them where employment is available and wages are high. Thus a sizeable part of the labour force cutting sugar cane in north-west Argentina comes from adjacent parts of Bolivia where wages are low and employment opportunities more restricted. Many Bolivian seasonal migrants remained and subsequently moved to the larger and more varied source of employment represented by Buenos Aires. Similar migration occurred within the past fifty years from northern Colombia to Venezuela in association with petroleum extraction and the development of urban and industrial services associated with it, both in the early petroleum centres around Lake Maracaibo and later to the Caracas region. Such population movements are commonplace in Latin America.

A further category of movement involving smaller numbers of people is migration in search of new land. Many frontier areas in Latin America are settled by few people and where disparities exist in availability of land on either side of a frontier then people seek to acquire land where it is most easily available. Thus Brazilians acquire land in eastern Bolivia, Salvadoreans move to adjacent Honduras and native Indian Guatemalans have moved to settle the southern and western parts of Belize. Such movements pose a threat to the recipient country since all countries fear territorial losses when citizens of a neighbouring country populate land close to their border and frontier conflict during the present century has regularly taken place between Latin American nations.

A final category of international migration that is of contemporary importance is that to and from the USA and other industrialized countries. On the one hand this includes the migration of those with or wishing to acquire special skills through education, and on the other, workers prepared to sell their labour at a rate sufficiently below that prevailing to ensure rapid if illegal employment.

The migration of those who have reached the top of the educational ladder – widely referred to as a 'brain drain' – is important because it represents a loss of valuable national human resources – for, even if some such migrants return, many stay or move to other countries. It also represents a waste of scarce national educational resources since highly educated people who leave to use their skills overseas have notionally deprived others of the chance to use similar finite educational resources during the course of their education at home. It creates a pattern of behaviour that others may imitate and thus a rolling snowball effect which may cause the numbers seeking to leave to increase. It also creates, through the influence of those who return, a pool of high-level knowledge which may not be relevant to the needs of national society and diminishes the innovative capacity of those with greatest intellectual ability. French-trained Ecuadorian engineers return with knowledge that may be appropriate to French engineering needs but which may be wholly inappropriate for specific national needs: worse, such experts may design training programmes for Latin Americans based on their own training rather than respond to the national situation of limited capital but abundant human resources.

Increasingly the capacity of Latin American institutions to offer education to their own people is expanding but creating a Latin Americanization of the migration of the best educated whereby Brazil, Argentina or Mexico increasingly attract the cream of Latin American scholars and technocrats both for education and for work to the detriment of their own countries. The formidable language barrier that is associated with study outside Latin America further encourages young people to study within their own continent and the increasing ignorance and misunderstanding of Latin America shown by Anglo-Americans and European governments further directs the migration to Latin American centres of perceived excellence.

The migration of less-skilled Latin Americans, largely to the USA, even though some have gone to the Persian Gulf states, is a phenomenon that has most affected Mexico although it has, in very specific contexts, also affected both Cubans and Haitians. The migration of Mexicans to the USA will be referred to later but it represents a major possibility for earning to tens of thousands of Mexicans at many different economic levels. The great economic differences on either side of the border mean that Mexicans can readily find casual and short-term employment in the USA which enables them to save money to send or take home and to finance illegally crossing the border and transportation in search of employment. This form of migration is a real alternative to internal migration within Mexico and in a variety of ways decreases pressure on Mexican resources, provides a large amount of foreign capital and benefits many people besides the migrants through the spending of migrant earnings within Mexico. It has negative effects also, in the cultural dislocation felt by migrants forced to travel long distances and live in an alien and frequently hostile cultural environment, even though many migrants live in migrant communities, often made up of people from their own part of Mexico. They acquire, at least in part, a life-style that is unsuited to rural Mexico to which many eventually return and thereby creates dissatisfaction which is not easily alleviated except through further periods of migration.

Conclusion

A rapidly developing continent freed from one form of colonialism only 160 years ago has increased its population by natural demographic increase and by the willing acceptance of large numbers of people from other continents. The consequence of these population movements has been to create not only a very culturally diverse environment but also to stimulate the use of many different cultural traditions in the search for well-being. The predominance of the capitalist economic system and the division of society along lines that were determined largely by differences in access to resources has also facilitated the conformity of people of these many cultural origins to patterns of society that are well known in western Europe and Anglo-America.

The diversity of cross-border migratory movement during the present century has reflected the levels of economic growth of different nations as well as regions within nations. Since such movements are often well-documented and recorded by national census data, migration may be used as a mirror of the geography of social and economic change and is therefore worth more detailed examination at a national and regional level.

Migration in the rural world

Migration is a part of the life-cycle of every person. Children migrate in association with parents, sometimes in order to be educated, on marriage and when seeking a living. Population movements have always been part of Latin American rural life, whether seasonal movement, in response to war, famine or political reorganization and there is little evidence that overall mobility changed greatly either during the Colonial Period or in the present century. What merits special attention is the composition and direction of migration. Migration at any given moment mirrors stresses in society and in different regions. Aggregation of the migratory behaviour of individuals in households, of households in communities or regions and nations gives some measure of certain aspects of population mobility and thus a glimpse of the pattern of social stresses. The nature of what is referred to as migration must be accompanied by a clear definition of what is meant. Here we mean migration to refer to any change of residence that takes place for more than a few days, whether or not the migrant is accompanied by other members of the household. Census definitions such as habitual residence may alternatively be used and it is important to realize that for every change in definition a different measure of human mobility is being taken. Another term frequently used in migration studies – permanent migration – deserves discussion. Migration permanency is a very elusive concept. Long-term Mexican migrants to Chicago may refer to their migration as temporary even after twenty-five years' residence because they perceive their home village in Mexico as the place to which they will ultimately return. If they die in Chicago, can their migration to the USA then be seen as permanent because of the accident of their death there? Some forms of migration – journey to work, holiday, marketing – may be cyclic and thus clearly temporary but the the use of a word qualifying migration which refers to probability of future mobility is hazardous and thus best avoided.

In this chapter we refer to migration in rural and urban worlds which implies a neat dichotomy between places which have clearly recognizable characteristics. Rural places contain people and households whose livelihood is largely derived from farming and livestock, while urban people are engaged in manufacturing, the distribution of goods and the provision of services. Rural people live in an environment in which households live in small agglomerations of ten to a hundred dwellings or in dispersed homesteads; urban people live in an environment comprising closely packed dwellings and other buildings where many thousands of households live in close proximity.

Such a neat division of human environments is inadequate with respect to migration simply because human mobility is such that urbanites seek rural residences, while some farmers choose to live in towns. Many households are involved in different ways of getting a living which encompass urban and rural environments for different periods of time and at different seasons of the year. Many Latin American rural families seek to urbanize their environment even though the whole household is predominantly occupied in farming and farming-related tasks. It is therefore unrealistic to suggest that people can be assigned to two mutually exclusive categories – urban and rural. Thus a division of mobility into that experienced by rural and urban people is inherently artificial and merely a device by which a series of mobility patterns can be described to enable some conclusions to be drawn about the extent to which mobility does provide an insight into the ways people use different environmental resources and into the pressures on people from the political, social and economic environment of which they are part.

Migration theory and Latin American experience

The examination of the state of knowledge about migration in Latin America must make reference to the substantial body of theory that exists regarding population mobility and of the alternative theoretical frameworks within which migration can be discussed. Much of migration theory is based upon emphasis on the economic motives underlying migration which are most readily quantifiable and which do provide a powerful if partial explanation of migration. People move towards areas where wage levels are high (cities) and where quality of life, expressed by housing availability, and the degree of provision of health and education is most satisfactory. Similarly, as we have already suggested, migration between countries is predominantly from poor countries towards richer countries.

There is also a group of migration theories which state that most migrants move short distances – in part a response to the Principle of Least Effort – and that the larger size of a populated centre the greater the attraction that it exerts on migrants. Important fundamental work on migration has been carried out in Sweden by Hägerstrand (1957) and, more recently by Carlstein (1982), which demonstrates the patterns of mobility associated with different phases of the human life cycle and with different roles of individuals within households.

Much of the reporting on research regarding migration in Latin America has been very much concerned with describing migration rather than attempting to relate the phenomenon in one area with a larger existing body of theory. Likewise few serious efforts have been made to generate theoretical statements about the general characteristics of migration. Some comments have already been made about the need for careful definition of the terms used in discussing migration but further care is needed in recognizing the level of explanation that is being sought. When a migratory event occurs the causes are complex and cannot be subsumed by referring to abstract situations such as 'population pressure' or even soil erosion. It is necessary to know the whole context in which migration is occurring, the extent to which the decision is made by the individual or the group and the length of time over which the decision has been considered. When interviewing a migrant she may explain that she moved to the town after a big row with her husband; but after talking further it emerges that she moved also to enable her daughter to attend a better secondary school, that she moved to live with an aunt who needed help in her shop, and that her two cows had died of foot-and-mouth disease and she had no alternative form of income. Thus a bundle of possible explanations of migration emerge without it being clear which, if any, can be identified as the 'real' cause.

The high level changes that are occurring in national and local societies may also be related to migration and various writers have pointed out that it is in the interest of larger-scale industry that there should be a plentiful supply of labour, to help keep wages down and to facilitate the maintenance of a rapid turn-over of labour in response to a changing economic situation. The move from labour-intensive to capital-intensive industry – including agriculture – also affects the need for people to move about to seek a living. Thus any analysis of migration should take into account the changes in society at large as well as those changes at a local, community and household level.

In this analysis of population mobility in Latin America we shall seek to report the situations in which migration is taking place and emphasize the impact of migration on both sending and receiving areas. We shall also try to show the extent to which migration is part of the human geography of rural and urban Latin America and the degree to which what is happening there accords or not with existing theory on human mobility.

Migration from the countryside

Migration from the countryside is commonplace throughout Latin America and has probably been so for several centuries. While some communities sought refuge from exploitation by minimizing their contacts with the outside, threatening Hispanic world, others could not and did not avoid such contact and sold goods and services, provided labour and produce as tribute in towns and cities and regularly traded and worked in ecological zones both different and far removed from their own. During the present century, population in many areas had finally recovered from the demographic catastrophe that followed the arrival of the Europeans in the fifteenth century with their strange and deadly diseases; but during the second half of this century population has actually declined or is stagnant in many parts of Latin America. This situation is demonstrated for the five central and northern Andean republics in Fig. 9.1. Detailed studies in Colombia show that a declining population was characteristic of large areas of the highlands in the intercensal period 1964–73 including 47 per cent of all *municipios*. In Ecuador, over a similar period, population decrease at a parish level occurred in many zones scattered the length of the Andean area but in some areas, such as Bolívar Province on the western side of the central Andes, a majority of parishes have experienced recent population decrease, much of which is assumed to have been the result of net loss of population through migration. The provisional 1982 Ecuadorean census data suggest that this situation has continued. The areas of population loss are predominantly in high-density non-Indian rural settlements. Data from Peru, mapped at a provincial level, show areas of population decline also in the more densely populated parts of the mountains. Data from both Bolivia and Venezuela show broadly similar patterns although the data base is not directly comparable. One may conclude however that extensive emigration is widespread. A detailed analysis of emigration in Argentina by Wilkie showed that this phenomenon was equally true of a more developed and industrialized country such as Argentina (Wilkie 1980).

The causes of this migration are complex but foremost among them for many households is the inability of rural levels of living to be maintained as the cost of inputs into agriculture (credit, machinery, fertilizers, seeds, agrochemicals) increase more rapidly than do prices of farm produce. Likewise the low esteem in which farming and rural life is held by national élites encourages rural people to seek a non-farming future for themselves and/or their children. Natural hazards such as droughts, floods, and soil impoverishment and erosion are all important in some places but need to be seen in the context of the broad range of problems facing farmers, outlined in Chapter 8.

The consequences of this migration for the sending communities are not all negative. In cases where migrants are absent for long periods they send money home to support non-migrants. This money is spent locally, stimulates commerce and may even stimulate investment in agriculture. The land of migrants, if not farmed by remaining family members, is rented (or sold) to non-migrants, thus

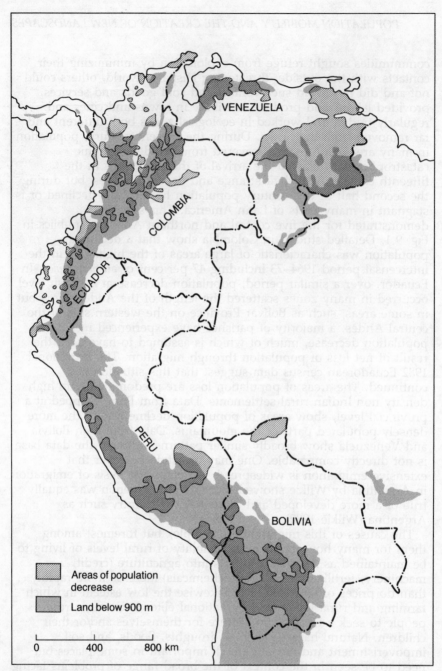

Fig. 9.1 Major areas of rural emigration in highland South America.
Notes on Sources: Bolivia: Data from 1976 Census, provisional results.
Provinces with an increase in population 1950–76 of less than 1 per cent;
Peru: Provinces with no change or a population decrease 1961–72 (Census of
Peru); **Ecuador**: Census of 1974. Parishes with a decrease in population
1962–74; **Colombia**: Municipios with a population decrease 1964–73;
Venezuela: States with a rural population decrease forecast 1980–2000 and
with negative migration balance 1961–71

Plate 14   Empty house, Solano, southern Ecuador, 1973. As more and more people leave for long periods, the countryside shows more signs of being deserted though migrants barricade doorways to show that they mean to return. (Photo: D. A. Preston)

increasing their access to resources. Returning migrants may come back with money and new ideas of ways to make a living including some which may be relevant to farming. In many cases the communities from which migrants have gone had little land per household and have for a long time been dependent on a variety of non-farming sources of income, such as carpentry, hat-weaving, rope-making. There, the money and ideas that result from migration go into non-farming activities rather than into farming which is seen as the source of food rather than as a basic source of goods to sell. In a successful savings and loan co-operative in a village in highland Ecuador which had experienced much emigration, only 22 per cent of the loans approved in 1975 were for crops or livestock, most were for home improvements and commercial ventures.

Somewhat similar findings come from a study of Mexican migrants to the USA from the State of Jalisco (Cornelius 1979). Migrants on average sent home about $200 a month but were also able to save another $300 or more to bring home with them when they returned. The most frequent investment was in land for a house or for farming but the most successful often attempted to start a small business. A major effect of migration to the USA is to stimulate the migrants to wish to acquire a broader range of consumer goods and to participate in modern consumer society. However unusual the

Plate 15 Dancing group at Iquiaca Fiesta, Bolivia, 1970. Migrants from this community living in La Paz form a dance group for the community fiesta but their costume of dark suits and ties and *chola* skirts and shawls demonstrate their urbanity. (Photo: D. A. Preston)

USA may be as a migration destination in Latin America as a whole, a variety of other studies of rural emigration towards major cities suggest very similar effects on migrants. Not all rural migrants head for the cities: our study in Ecuador in 1975–76 showed that, while 66 per cent of the children of households visited who had left the parish lived in urban places, only 52 per cent of the returned migrants interviewed who had migrated within the previous two years, had been to towns and the proportion of returned migrants who had migrated to urban areas in the 1950s was only 26 per cent. Urban destinations are thus a relatively recent preference for some rural migrants. Indeed, using 1974 census data for Ecuador and taking 'place of previous residence' as a migration definition we find that about half of rural migrants move to other rural destinations. Even when local intraprovincial migration is filtered out, over a third of rural migrants head for other rural areas.

Our own evidence from case studies in Ecuador concerning the amount of migration from those rural areas experiencing considerable population loss suggests perhaps 60 per cent of the generation of children born in 1950 now live elsewhere. A study in Jalisco in Mexico found that in one village 75 per cent of sons or brothers of sample farmers lived elsewhere, while only 35 per cent from a neighbouring village had left; a detailed study in Oaxaca showed that 60 per cent of the sample area had migration experience while Wilkie's study in a village in Entre Rios in Argentina showed that 45 per cent of the population had migrated since 1954 (Wilkie 1980).

Rural emigration has frequently been held to be detrimental to rural areas because it leads to the loss of the 'best' people and when it continues for several generations those that remain are the oldest, most conservative and least imaginative. There is abundant evidence to show that the characteristics of migrants are very similar to those of innovators which might suggest that little innovation might be expected in areas with a lot of emigration. However, one of the basic principles of migration is that every outflow generates a corresponding inflow; certainly a proportion of the migrants from the rural areas we have studied return to their home. Since much migration is short-stay, particularly in Indian areas, then any community is likely to contain a number of returned migrants. Thus the community is not denuded of all those with a broad-world view, with the capacity to innovate and the willingness to take risks for the possibility of larger gain, for migrants return with money and experience gained away and the community may thereby gain from their knowledge. Furthermore, while migrants are frequently more educated than non-migrants, if their education has not fitted them for livelihoods that can be carried on in their home area, then their knowledge and ability is not crucial to their community. If a butcher or carpenter migrates their village will only suffer from their going if there is no other butcher or carpenter left. Rural emigration does not thus necessarily cause a loss of people in the labour force who are necessary to the maintenance of existing levels of living in the community. Furthermore whether such migrants do constitute the 'best' people depends on *post facto* value judgements that may be hard to verify.

The decline in population affects specific parts of rural areas in different ways; the absence of migrant males, for instance, affects specific tasks as varied as ploughing fallow land and occupying positions of authority in the community, for both of which male migrants are sometimes called home. The loss of younger people naturally affects reproduction rates although some migrants bear children while away and yet leave them in the village while they work elsewhere (particularly women employed in domestic service). The most profound consequence of migration is the exposure of many people and their kinsfolk to other ecological zones, to different ways of making a living and to other social and cultural values which they would not otherwise have known. A consequence of this increased circulation of people between towns, cities and rural areas is a need for cash with which to purchase what is newly felt to be essential. In the areas we have studied this did not lead to increased commercial farming although this has occurred in some places. It has led to a greater degree of social and political awareness and to a gradual proletarianization as people become involved in ways of production other than farming and handicrafts. Migration does not necessarily lead to the decline of rural areas. It leads to a redefinition of the space in which household livelihood is maintained. A rural-based household may become, as a result of migration, part of a complex multi-household livelihood strategy where co-operation between kin and friends in different locations enables economic and personal crises to be overcome and security as well as profit is maximized.

The search for new land

A completely separate aspect of population mobility in relation to rural areas is that concerned with new land settlement. This is also the aspect which has traditionally received most attention from geographers for it relates to assessment of the agricultural potential of the physical environment and the process of new land settlement in Latin America has been important and well-documented for the past 150 years. In the past decade the settlement of vast areas of the Amazon basin associated with a major road building programme embarked on by the Brazilian military government has impressed many and also given rise to fears for the ecological future of this major region. The recent establishment of farms in other lowland humid tropical lands may also have led to major environmental changes, the consequences of which are still being experienced.

The search for new land has been associated with the changing needs of rural populations since prehistory. The movement of the Mayans from the limestone lowlands of Yucatán to the hills on the Central American cordillera, and of the Incas to settle and/or subjugate other highland and lowland areas are examples of new land settlement in the past. It is now widely recognized that many Andean peoples had access to land in different ecological zones which included both high altitude grasslands and sub-tropical lowlands. It is the search for suitable lowland farmland that is the most important cause of most new land settlement in Latin America today. What concerns us here, in the context of population mobility, is the nature of the search for new lands, the origins of migrants, the nature of their destinations and the changes in the human geography that result from such migration.

The most important cause of the search for new land is that the would-be colonist feels that settlement in such an area will allow more freedom and opportunity both for the adults and for any children that they have or expect to have. For colonists from urban backgrounds this may be because of high rents and the fierce competition for employment or for clients in big cities; for rural people it may be because of land shortage, recurrent climatic crisis or low returns from domestic craftwork. Others are escaping political strife (as in Colombia during the *violencia* of the 1950s), or the police seeking them for serious crimes or parents-in-law for carrying off their daughter, even if not against her will. Some feel that a new social order will be created on the frontier where people will be judged by what they achieve rather than by their parentage.

The geographical and other origins of colonists is varied. In some areas the majority of colonists are from rural areas, in others they are largely from small towns and even cities but in many cases they have at some time had experience of farming. Not all 'colonists' intend to become farmers, for some acquire land, farm it for a year and rent it to another family while they establish themselves as merchants or shopkeepers, only visiting the farm once a month or so and taking little part in its management. Many colonists come from areas where land is in short supply, typically from the highlands (in much of Central and Andean America) or from other areas where life is hard. Few colonists come without previous experience of the lowlands and many have migrated there as labourers on many occasions and thereby gained knowledge about the area and its possibilities. Many have kinsfolk already farming who can give guidance and, in some cases, colonists are moving from another newly settled area as the land there becomes less productive. A recent study by Connie Weil of one such area in the Chapare zone of Bolivia showed that four-fifths of the farmers had moved from another part of the same colonization zone although most had been born in the adjacent highlands (Weil 1983).

Moreover, colonists often retain land in their home community in the highlands, partly as an insurance against failure but also because they intend to travel regularly between the two areas. In Weil's study colonists spent as much as 25 per cent of their time cultivating their other land whether in the highlands or in an older colonization zone. In Mexico and Guatemala highland families regularly migrate to work in the Gulf lowlands and some eventually obtain land there and cultivate crops in both zones.

New settlers come to the new lands in one of two ways. A minority come as settlers on directed colonization schemes. They are allocated a parcel of land, given advice on what to grow, loaned money to get over the initial period before the first harvest and generally given help to make a success of their new life. The majority are spontaneous colonists who have found their own land, either bought it from the previous owner or claimed it from the government's land office. They receive no help and there is no effective scrutiny of their past to attempt to determine whether or not they are suitable colonists. Large sums of money, often from international agencies, are spent on the former group – as much as $10,000 per family in some cases – to ensure success and yet such schemes have a high turnover of settlers, more of them default on bank loans than the self-selected spontaneous colonists who tend to persist longer in their efforts to succeed. One of the reasons for this disparity in levels of success is that the government agencies directing colonization select poor sites for farmers and give little advice which in any case may be inappropriate. Their criteria for selecting farmers are poorly chosen, and, perhaps most important of all, the directed colonists are tempted or forced to place too great a reliance on the managers of the project rather than follow their own judgement. Recent Mexican colonization experience has laid great emphasis on direction from above in the Gulf coast schemes, such that colonists feel themselves to be mere government employees. Other work in the Amazon basin suggests that new colonists pay insufficient attention to the farming methods of the few indigenous or long-standing farmers who have, through a long process of trial and error, evolved a stable and productive farming system.

The most important factors in the success of colonists seem to be having some financial resources to support their enterprise. Those colonists with most resources or most land elsewhere were frequently those who enjoyed most success. They buy up the land of the poorer colonists and acquire sizeable holdings which, in the Santa Cruz area of Bolivia, facilitate the transfer to cattle rearing which is often the second stage of land use following forest clearance. There is no evidence to support the idea that the frontier really does offer a new social order, for the better-off become the new élite and employ the poor to work their land, often employing the former owner of land that they have bought, as sharecropper or tenant on the land they formerly owned.

Conclusions

Migration is clearly a process that affects the majority of Latin Americans in both rural and urban areas, but the areas affected by different types of migration vary from time to time. Migration has been a major part of the process by which new land has been settled, both after the Inca conquest, and during the Colonial Period. It was particularly important during the nineteenth century. In recent decades, the growth of new urban centres, especially those associated with grandiose new developments such as those at Brasília and Cuidad Guayana, has been in a large measure the result of large-scale immigration from other urban areas and from the surrounding rural areas.

The most important conclusion that emerges from this review is that investigation of a single aspect of the migration process such as migration destination or flow is of largely descriptive value. The broad pattern of migration through space and over time must be seen as a reflection of the major changes in society. The most important contemporary change is the increasing demand by marginal households to have more consumer goods and access to the symbols of participation in national society to a far greater extent than ever before. Thus Bolivian Altiplano peasants in 1981 were buying gas cookers, televisions and the more affluent were buying video-recorders where ten years previously cooking was exclusively by dried llama dung and wood, or kerosene among the more affluent, and television was only seen in La Paz when visiting. Such innovations were paid for by wages earned when away from home and by increasing the amount of farm produce sold. These changes and the migration associated with them can most convincingly be accounted for by identifying the reasons behind the demand for more cash as well as the range of means by which such capital is acquired.

Latin America is unlike many other areas of the Third World in relation to migration in several respects. The continent is highly urbanized, the population relatively concentrated around the periphery, and already experiences a level of living superior to elsewhere in the Third World. Oil wealth and revenue from the export of other primary produce has facilitated the extensive development of empty lands within the continent stimulating further migration to rural zones. The basic relationship between migration and social change is, however, common to people throughout the world and is not peculiar to Latin America.

Migration and urban places

Although we emphasized previously the arbitrary nature of considering migration in relation to urban and rural places, migration in relation to urban places is particularly worthy of special attention. As an increasing proportion of Latin America's population is urban-based then migration to, from, within and between urban places will inevitably become a more important component of migration. Furthermore, since urban populations are growing rapidly in absolute terms and also experiencing rapid socio-economic change then the characteristics and consequences of migration involving urban locations is of particular importance. Since a widely held view of urban growth in Latin America is that it is largely a consequence of migrants swelling urban population, it is necessary to indicate the variety of reasons for urban growth other than immigration and to identify the specific role that migrants play in the dynamic urban geography of Latin American urban centres.

Migration and urban growth

The growth of the urban places of Latin America is not solely the consequence of the arrival of large numbers of migrants. In many countries the commonly accepted belief that large cities grow faster than smaller centres is true and there is a variety of evidence that supports the relevance of the basic theory of migration that migration flows are proportional to the size of the receiving area. This is by no means universal, however (see Table 9.1). For instance, as a consequence of deliberate government policy in Cuba since 1959, population increase has occurred to a far greater extent than formerly in smaller cities and in the countryside. The 1981 Cuban census showed that the city of Havana with a population of 1,900,000 people had only increased its population by 0.7 per cent per year, while the two next largest cities had grown by 2.2 per cent

251

Table 9.1 Annual increase in
population (1960–70) in Latin
American cities

City size	% Population increase
1,000,000+	5.05
500,000–999,999	2.35
100,000–499,999	7.08
50,000–99,999	4.07
20,000–49,999	4.45
Overall increase	5.03

Source: Wilkie and Haber (1983)

and the four cities with between 100,000 and 200,000 had grown by 3.0 per cent per year (1970–81) (See Table 9.2). Likewise, although in a completely different political context, recent Ecuadorian census data show that some of the smaller towns are the fastest-growing urban centres in the country. Population change at a micro-regional level tends to be a sensitive reflection of the pattern of local economic and political development. A feature of a number of Indian areas of Ecuador and Bolivia has been the atrophy of small regional centres which were dominated by non-Indians and where racial discrimination was widely practised. In newly created urban centres (especially in highland Bolivia) and in urban centres in more highly commercialized rural areas, by contrast, the increase of the urban population has been more rapid.

Table 9.2 Urban population change in
Cuba 1970–81

City size	Number of cities	% annual increase
More than 1 million	1	0.7
250,000–1 million	2	2.2
100,000–250,000	4	3.0
National increase		1.6

Source: *Granma* (1981.)

The growth of urban population is the consequence of the high rate of natural increase of the population of the largest cities and also the increase which is the result of in-migrants being more numerous than out-migrants. A group of factors is associated with the high rate of natural increase of urban population. The age structure of the population shows that a larger proportion of the population is capable of reproduction at least in part because migrants are predominantly in the 15–35 age group. Therefore, any population containing a substantial proportion of recent migrants is likely to contain more than the average number of people in this most fecund age group. Furthermore, the level of health care facilities is higher in urban centres, especially in the metropolitan centres. The infant mortality rate is therefore lower and this further contributes to the increase in population.

Evidence concerning the proportion of the increase in big city population that can be attributed to immigration is conflicting. In an analysis of Venezuelan data for the 1940s, 71 per cent of the population growth was the result of immigration but a similar analysis of Mexican data attributed only 42 per cent of the increase to immigration. The importance of migration has increased during the last twenty-five years but even so it is unwise to assume that urban growth is largely the consequence of immigration: it is the result of a series of interacting forces.

Characteristics of migration to towns and cities

Migration has a pronounced impact on the social, economic and political complexion of both the sending and receiving areas for migrants are not uniformly representative of all social groups. Likewise, if they differ markedly in a number of respects from the host population, friction may result and also the composition of the urban population will change. Although most of what has been written about urbanward migrants refers to cities, it is also worthy of note that urbanward migration occurs to almost any urban centre whether it be a small village serving a numerous and scattered rural population or a regional capital, centre of a zone containing a booming plantation economy, a populous highland Indian area and a network of small market centres.

By way of illustration migration patterns for a part of the Ecuadorian Andes can be indicated. The parish centre of Quilanga is a large village with a population of 700 people but the social and economic centre of a rural area with a population of some 3,600 people. The parish centre attracts migrants, apart from government employees and merchants from outside the parish, primarily from the outlying rural areas who seek what, in their words, is the more 'civilized' environment of the little town. There could be found in Quilanga, two single-sex primary schools with a teacher for each grade, a secondary school, a medical centre and a priest quite apart from a number of shops selling a wide range of goods, several buses each day to the provincial capital of Loja and all the social intercourse possible in a small town at the end of the day and at weekends even though there was no weekly market.

Some of those who live in Quilanga town, however, find it

depressing and intend to leave. They yearn for brighter lights, for the chance of jobs other than related to farming. They want a secondary school that offers the complete secondary school curriculum instead of just the first part, they want to be able to buy a wider range of goods more cheaply, to have a home with electricity and running water and to be able to feel less isolated from the trends of life reported by the media. They envy kinsfolk in Loja with television and the young people want to be able to have the opportunity of going to the cinema or buying a drink in a bar where their behaviour will not automatically be noted and perhaps reported to their parents. So some of the small townsfolk go to Loja, a city of 66,000 people, where there are half a dozen buses a day to Guayaquil, the coastal metropolis and to Quito the capital and highland urban centre.

The people of Loja, for their part, are conscious of the small size of their city, that it is inaccessible except by a long, and relatively expensive road journey from the big cities and being a few hours from a recently opened road to the jungle is no compensation for the time and cost of getting to Guayaquil, a million city. They hear from relatives and on the radio and television of the range of jobs available in the big city, of the better quality of schooling, for all the best teachers and doctors want jobs there and they are well aware that politicians are more readily influenced by petitions from citizens in the big city than in the distant country areas. A teachers' strike in Loja would be relegated to the inside page of a national newspaper, one in Guayaquil would be worthy of a spot on the television news and on the front page of the main daily paper.

Thus there is a current of migration up and down the settlement hierarchy as individuals and groups seek to satisfy their aspirations and migration becomes an important dynamic component of population change in many categories of urban centres.

It is axiomatic that migration flows contain a powerful element of momentum and many migrants are led by others from a similar area and with the same goals so that it is common that migrants from the same community or area will congregate in the same part of their urban destination. The extent to which migrants segregate themselves in particular geographical areas is much debated but in the period immediately after initial migration many of the migrants from any zone of origin are likely to be living in close proximity although as they acquire more knowledge about the town or city they move more independently. Where migrants from a particular area have tended to find similar sorts of employment, as, for example, flower-sellers in one Mexico City market have done, then residential propinquity may be maintained. Where there is a major cultural difference between migrants and urban population then migrants are more likely to stay together for reasons of solidarity until they can communicate on equal terms with the population as a whole. Some migrants also remain in close proximity in their urban destinations for commercial purposes.

The network of links that develops both between migrants and between migrants and their home areas is important for economic survival and for social satisfaction. These links are not necessarily based on migrants living close together but there are frequent occasions when migrants from the same area meet together to play football, to plan political action on behalf of the home community or to organize their contribution to major festivities either in the city (for a national celebration) or in the home community for a festival. These links enable migrants to weather periodic personal crises and the flows of goods are in both directions between newly arrived and long-standing migrants and between migrants and stay-at-homes.

Migrants are frequently different from the urban population as a whole. We have already indicated that they are frequently younger than the average urban dweller, they often have a different range of skills and their level of education is lower than the urban population as a whole. Yet various studies show that migrants are not relatively unsuccessful at finding jobs and making their way in the larger town. Although they may suffer discrimination if they belong to a particular ethnic group, they are not necessarily handicapped by reason of being recent migrants. Migrants too are not from any particular social stratum. Although many migrants from rural areas are relatively poor they often maintain resources at home. They seldom sell any land that they may own and better-off landowners as well as poorer peasants migrate to urban centres.

Migration from and within urban areas

Migration from urban areas is not necessarily related to any deterioration of the quality of urban life. It does, however, increase at times of economic depression and decrease when jobs are plentiful and wages adequate. In addition there is an element of outward migration to establish businesses in smaller towns or villages and to use the information networks that have developed during years of urban living. Other urban out-migrants have tired of city life, of the high cost of living, especially housing, of crime and violence, of extensive atmospheric pollution and of the high pressure of city life. It is they who return to their village to set up a shop, to buy land or livestock and farm. It is also they who go to colonization areas. They thereby act as disseminators of urban values and life styles in far-off, non-urban areas.

Another increasingly common form of urban emigration is that of suburbanization and the creation of dormitory dwellings in the countryside on the periphery of the biggest Latin American cities in exactly the same way and for similar reasons as such urbanites seek rural homes in the industrialized nations. Caracas, Bogotá, Mexico City and Lima all have areas where the more affluent of the city chose to live up to an hour or more's travel from the city centre. Such areas are not necessarily accessible by public transport but the presence of commuters gradually transforms the rural towns or village as new residents seek and obtain better municipal services. Such suburbanization has been little studied in Latin American cities but to a visitor the spread of city folk into the rural areas is easily noted and is an important element in the changing urban geography of big cities.

A further feature of urban growth which is likewise associated with some of the largest Latin American cities is the rapid spread of the poorest households into the far periphery of the city. This inverts the customary social geography of Latin American cities where the poor settle the less desirable sites as close to the centre as possible – swamplands, hillsides, etc – and it seems to have been the result of repressive policies of city and national governments that have placed severe restrictions on the growth of informal, semi-legal, low-income housing within the city periphery and thereby forced the poor to migrate from the city and make their homes in undeveloped areas beyond the city limits. Thus entering Caracas from the coast the first sign of the city is shanty-towns clustered on steep hillsides several kilometres from the paved roads of the suburbs of the city. Several writers have documented similar phenomena in São Paulo where vacant land nearer the city centre which would, in a different political climate, be settled on by squatters, is held by land speculators; the poor are forced out to the very margins of one of the most populous and sprawling cities in the hemisphere and thus are forced to travel for two to four hours daily to and from work from their homes on the very fringe of the city. This relatively recent phenomenon has received little serious investigation but is certainly a striking feature of some Latin American cities.

The new landscapes

The human and the physical environment of Latin America is continuously changing in response to new ways of making a living being followed by its inhabitants, by the changes that long-term modifications in the physical environment stimulate in the resident population and by the movement of people into and out of different parts of the continent.

The new landscapes that are a consequence of these changes offer a complex assemblage of evidence of the ways in which people subsist and of the environments that they modify and create. It is the study of this assemblage of evidence that is the particular concern of human geography and it is the identification of landscapes and of patterns of change in space and in time that has long been a major contribution that human geography has made in Latin America.

Population migration has been important in influencing the fashioning of new landscapes, both physical and human. Although when the Europeans came to the New World what is now Latin America was populated by large numbers of people, in many areas, particularly the western highlands, the decline in native population was more important than the immigration of the Europeans in modifying rural landscapes. The new lords of the land did bring new animals and new crops that transformed large areas. The new urban centres represented a new man-made environment on a scale not previously known and many of the urban centres were peopled by indigenous migrants from rural areas – Indians needed as servants and workers, skilled craftspeople to manage industrial processes – and by the new aristocracy from across the ocean. The form and style of the urban environment was largely fashioned by the Spaniards and Portuguese according to their conception of urban plan and all but the most humble homes conformed to Iberian architectural forms even though all workers and many craftspeople were natives.

The major changes to the urban environment during the past hundred years also are related to growth of urban centres associated in part with the creation of new homes for migrants first in the period of massive European immigration to the southern countries in the last quarter of the nineteenth century and more recently the rapid growth of cities associated at least in part with migration from other towns and the countryside. These new landscapes, vast in extent, convey something of the origins of the migrants not so much in the style of the houses as in the range of activities that occur and, visually, in the names of the businesses and the games and music played by the people.

The physical environment, the vegetation, farmland uses and the characteristics of the slopes and of the hydrological cycle all reflect the use which people have made of the land. Migrants are the prime architects of the living environment in the new agricultural areas of the humid tropics and the new systems of intensive farming being developed in many parts of Latin America are likewise foreign in origin and frequently use a labour force comprising people who themselves are migrants. For example, the carnation flower-workers of the basin of Bogotá are largely young women from the poorer city districts. The banana plantations of the Caribbean coasts are directed by North Americans, and worked by highland migrants and local peasant farmers. In each case they are people willing to migrate some distance to obtain well-paid short-term work at times when their work is less needed at home.

Although much change is initiated from within households and communities either migration experience or migration itself is profoundly important in stimulating change which eventually leads to the transformation of rural and urban habitats. Change cannot be understood merely by identifying the migratory processes that may be associated with it. Migration is worthy of study because it helps to reveal many of the tensions of regional and national society. Migration is the direct consequence of human decisions. It arises from needs which themselves reveal much about individual and collective evaluation of social position. Thus urbanward migration as a major social trend is the logical consequence of individual judgements that the chances of improving the quality of life are better in the city destination than where they came from. It also reveals a preference for the values and life-style associated with urban living at whatever position on the social scale.

Further reading

Barbira-Scazzochio F (ed). (1980) *Land, People and Planning in Contemporary Amazonia*. Occasional Publication No. 3. Centre for Latin American Studies, University of Cambridge, Cambridge. (Collection of essays about many important issues concerning the Amazon basin and general issues relating to colonization.)

Butterworth D and **Chance J K** (1981) *Latin American Urbanization*. CUP, Cambridge. (A useful and thoughtful review of a wide range of literature concerning urbanward migration and change within growing cities with an anthropological bias.)

Carlstein T (1982) *Time, Resources, Society and Ecology*. Studies in Human Geography No. 49. Department of Geography, University of Lund, Lund.

Cornelius W A (1979) 'Migration to the US. The view from rural communities', *Development Digest* **17**, 90–101. (Good short summary of broad-ranging research on Mexican migration northwards.)

Crossley J C (1983) 'The River Plate republics', in **Blakemore H** and **Smith C T** (eds.), *Latin America: Geographical Perspectives*, pp. 383–455, Methuen, London.

Faron L C (1967) 'A history of agricultural production and local organization in the Chancay valley, Peru', in **Steward J H** (ed.), *Contemporary Change in Traditional Societies*, pp. 229–94. University of Illinois Press, Urbana. (An excellent detailed and well-illustrated account of the history of settlement and associated migration in a valley on the Peruvian coast from the Colonial Period until the present century.)

Granma 1981, Nov. 8.

Hägerstrand T (1957) 'Migration and areas. Survey of a sample of Swedish migration fields and hypothetical consideration of their genesis', in **Hannerberg D** (ed.), *Migration in Sweden*. pp. 27–158 University of Lund, Lund.

Kowarick L (1978) *The logic of disorder: capitalist expansion in the metropolitan area of São Paulo State*. Discussion Paper No. 202. Institute of Development Studies [at the University of Sussex], Brighton. (Good accounts of the nature and special characteristics of São Paulo's growth since 1964.)

Merrick T W, and **Graham D H** (1979) *Population and Economic Development in Brazil*. Johns Hopkins University Press, Baltimore.

Peek P and **Standing G** (eds.) (1982) *State Policies and Migration. Studies in Latin America and the Caribbean*. Croom Helm, London. (Good chapters on the regional and national issues in Cuba, Chile, Ecuador and Mexico with a socio-economic viewpoint.)

Preston D A (ed.) (1980) *Environment, Society and Rural Change in Latin America*. John Wiley & Sons, Chichester. (Contains various relevant essays on migration in Mexico, Argentina and Ecuador.)

Weil C (1983) 'Migration away from landholdings by Bolivian campesinos', *Geographical Review*, **73**, 182–97.

Wilkie R W (1980), 'Migration and population imbalance in the settlement hierarchy of Argentina', in **Preston D A** (ed.), *Environment, Society and Rural Change in Latin America*. pp. 157–84 John Wiley & Sons, Chichester.

Wilkie J W and **Haber S** (eds.) (1983) *Statistical Abstract of Latin America*, Vol. 22, Table 637, p. 91. UCLA, Latin American Centre Publications, Los Angeles.

Index